AS & A2 Physics

Exam Board: OCR A

Complete Revision and Practice

AS-Level Contents

A2-Level Contents

Editors:
Amy Boutal, Sarah Hilton, Alan Rix, Julie Wakeling, Sarah Williams

Contributors:
Tony Alldridge, Jane Cartwright, Peter Cecil, Peter Clarke, Mark A. Edwards, Barbara Mascetti, John Myers, Zoe Nye, Moira Steven, Andy Williams

Proofreaders:
Ian Francis, Glenn Rogers

Published by Coordination Group Publications Ltd.

ISBN: 978 1 84762 419 2

Data used to construct stopping distance diagram on page 26 from the Highway Code. Reproduced under the terms of the Click-Use Licence.

With thanks to Jan Greenway for the copyright research.

Groovy website: www.cgpbooks.co.uk
Jolly bits of clipart from CorelDRAW®
Printed by Elanders Hindson Ltd, Newcastle upon Tyne.

Based on the classic CGP style created by Richard Parsons.

AS-Level
Physics

Exam Board: OCR A

The Scientific Process

'How Science Works' is all about the scientific process — how we develop and test scientific ideas.
It's what scientists do all day, every day (well, except at coffee time — never come between a scientist and their coffee).

Scientists Come Up with **Theories** — Then **Test Them**...

Science tries to explain **how** and **why** things happen — it **answers questions**. It's all about seeking and gaining **knowledge** about the world around us. Scientists do this by **asking** questions and **suggesting** answers and then **testing** them, to see if they're correct — this is the **scientific process**.

1) **Ask** a question — make an **observation** and ask **why or how** it happens.
   E.g. what is the nature of light?

2) **Suggest** an answer, or part of an answer, by forming:

   • a **theory** (a possible **explanation** of the observations)
      e.g. light is a wave.

   • a **model** (a **simplified picture** of what's physically going on)

3) Make a **prediction** or **hypothesis** — a **specific testable statement**, based on the theory, about what will happen in a test situation.
   E.g. light should interfere and diffract.

4) Carry out a **test** — to provide **evidence** that will support the prediction, or help disprove it. E.g. Young's double-slit experiment.

The evidence supported Quentin's Theory of Flammable Burps.

A theory is only scientific if it can be tested.

...Then They **Tell** Everyone About Their **Results**...

The results are **published** — scientists need to let others know about their work. Scientists publish their results in **scientific journals**. These are just like normal magazines, only they contain **scientific reports** (called papers) instead of the latest celebrity gossip.

1) Scientific reports are similar to the **lab write-ups** you do in school. And just as a lab write-up is **reviewed** (marked) by your teacher, reports in scientific journals undergo **peer review** before they're published.

2) The report is sent out to **peers** — other scientists that are experts in the **same area**. They examine the data and results, and if they think that the conclusion is reasonable it's **published**. This makes sure that work published in scientific journals is of a **good standard**.

3) But peer review **can't guarantee** the science is **correct** — other scientists still need to **reproduce** it.

4) Sometimes **mistakes** are made and bad work is published. Peer review **isn't perfect** but it's probably the best way for scientists to self-regulate their work and to publish **quality reports**.

...Then **Other Scientists** Will **Test** the Theory Too

Other scientists read the published theories and results, and try to **test the theory** themselves. This involves:
• Repeating the **exact same experiments**.
• Using the theory to make **new predictions** and then testing them with **new experiments**.

If the **Evidence** Supports a Theory, It's **Accepted** — for Now

1) If all the experiments in all the world provide evidence to back it up, the theory is thought of as **scientific 'fact'** (for now).

2) But they never become **totally undisputable** fact. Scientific **breakthroughs or advances** could provide new ways to question and test the theory, which could lead to **new evidence** that **conflicts** with the current evidence. Then the testing starts all over again...

And this, my friend, is the **tentative nature of scientific knowledge** — it's always **changing** and **evolving**.

The Scientific Process

So scientists need evidence to back up their theories. They get it by carrying out experiments, and when that's not possible they carry out studies. But why bother with science at all? We want to know as much as possible so we can use it to try and improve our lives (and because we're nosey).

Evidence Comes From Controlled Lab Experiments...

1) Results from **controlled experiments** in **laboratories** are **great**.
2) A lab is the easiest place to **control variables** so that they're all **kept constant** (except for the one you're investigating).

> For example, finding the resistance of a piece of material by altering the voltage across the material and measuring the current flowing through it (see p. 50). All other variables need to be kept the same, e.g. the dimensions of the piece of material being tested, as they may also affect its resistance.

... That You can Draw Meaningful Conclusions From

1) You always need to make your experiments as **controlled** as possible so you can be confident that any effects you see are linked to the variable you're changing.
2) If you do find a relationship, you need to be careful what you conclude. You need to decide whether the effect you're seeing is **caused** by changing a variable, or whether the two are just **correlated**.

"Right Geoff, you can start the experiment now... I've stopped time..."

Society Makes Decisions Based on Scientific Evidence

1) Lots of scientific work eventually leads to **important discoveries** or breakthroughs that could **benefit humankind**.
2) These results are **used by society** (that's you, me and everyone else) to **make decisions** — about the way we live, what we eat, what we drive, etc.
3) All sections of society use scientific evidence to make decisions, e.g. politicians use it to devise policies and individuals use science to make decisions about their own lives.

Other factors can **influence** decisions about science or the way science is used:

Economic factors

- Society has to consider the **cost** of implementing changes based on scientific conclusions — e.g. the cost of reducing the UK's carbon emissions to limit the human contribution to **global warming**.
- Scientific research is often **expensive**. E.g. in areas such as astronomy, the Government has to **justify** spending money on a new telescope rather than pumping money into, say, the **NHS** or **schools**.

Social factors

- **Decisions** affect **people's lives** — e.g. when looking for a site to build a **nuclear power station**, you need to consider how it would affect the lives of the people in the **surrounding area**.

Environmental factors

- Many scientists suggest that building **wind farms** would be a **cheap** and **environmentally friendly** way to generate electricity in the future. But some people think that because **wind turbines** can **harm wildlife** such as birds and bats, other methods of generating electricity should be used.

So there you have it — how science works...

Hopefully these pages have given you a nice intro to how science works, e.g. what scientists do to provide you with 'facts'. You need to understand this, as you're expected to know how science works yourself — for the exam and for life.

Scalars and Vectors

Mechanics is one of those things that you either love or hate. I won't tell you which side of the fence I'm on.

Scalars Only Have Size, but Vectors Have Size and Direction

1) A **scalar** has **no direction** — it's **just an amount** of something, like the **mass** of a **sack of meaty dog food**.

2) A **vector** has magnitude (**size**) and **direction** — like the **speed and direction** of next door's **cat** running away.

3) **Force** and **velocity** are both **vectors** — you need to know **which way** they're going as well as **how big** they are.

4) Here are a few examples to get you started:

Scalars	Vectors
mass, temperature, time, length, speed, energy	displacement, force, velocity, acceleration, momentum

Adding Vectors Involves Pythagoras and Trigonometry

Adding two or more vectors is called finding the **resultant** of them.
You find the resultant of two vectors by drawing them '**tip-to-tail**'.

Example

Jemima goes for a walk. She walks 3 m North and 4 m East. She has walked 7 m but she isn't 7 m from her starting point. Find the magnitude and direction of her displacement.

First, draw the vectors **tip-to-tail**. Then draw a line from the **tail** of the first vector to the **tip** of the last vector to give the **resultant**:
Because the vectors are at right angles, you get the **magnitude** of the resultant using Pythagoras:

$R^2 = 3^2 + 4^2 = 25$
So $R = 5$ m

Jemima's 'displacement' gives her position <u>relative</u> to her starting point.

Now find the **bearing** of Jemima's new position from her original position.

You use the triangle again, but this time you need to use trigonometry. You know the opposite and the adjacent sides, so you need to use:
$\tan \theta = 4 / 3$

$\theta = 53.1°$ Trig's really useful in mechanics — so make sure you're completely okay with it. Remember SOH CAH TOA.

Jemima

Use the Same Method for Resultant Forces or Velocities

If the vectors aren't at right angles, you'll need to do a scale drawing.

Always start by drawing a diagram.

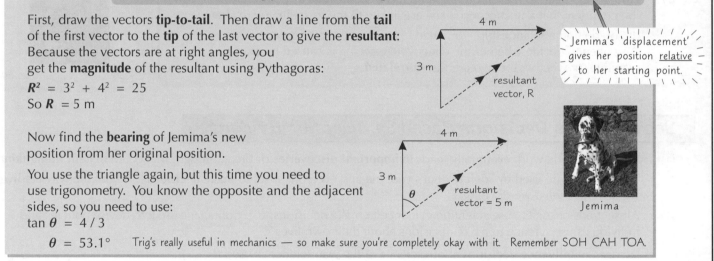

Example

2 N

add vectors
tip-to-tail

R

2 N

2 N

2 N

You know the resultant force is at 45° to the horizontal (since both forces are the same size).
So all you need to do is use Pythagoras:
$R^2 = 2^2 + 2^2 = 8$
which gives $R = 2.83$ N at 45° to the horizontal.

Don't forget to take the square root.

Example

8 ms⁻¹

add vectors
tip-to-tail

R

8 ms⁻¹

14 ms⁻¹

θ

14 ms⁻¹

Start with: $R^2 = 14^2 + 8^2 = 260$
so you get: $R = 16.1$ ms⁻¹.
Then: $\tan \theta = 8/14 = 0.5714$

$\theta = 29.7°$

Scalars and Vectors

Sometimes you have to do it backwards.

It's Useful to Split a **Vector** into **Horizontal** and **Vertical Components**

This is the opposite of finding the resultant — you start from the resultant vector and split it into two **components** at right angles to each other. You're basically **working backwards** from the examples on the other page.

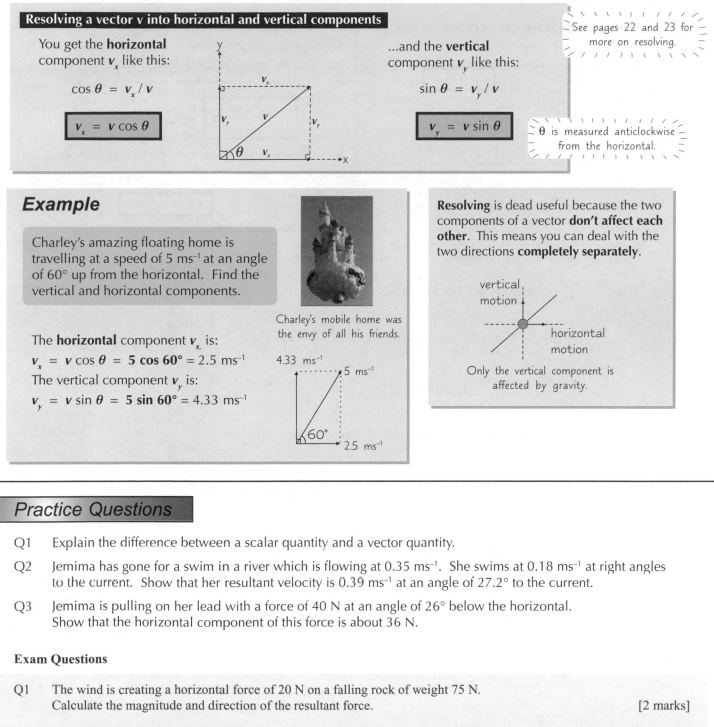

Resolving a vector **v** into horizontal and vertical components

You get the **horizontal** component v_x like this:

$$\cos \theta = v_x / v$$

$$\boxed{v_x = v \cos \theta}$$

...and the **vertical** component v_y like this:

$$\sin \theta = v_y / v$$

$$\boxed{v_y = v \sin \theta}$$

See pages 22 and 23 for more on resolving.

θ is measured anticlockwise from the horizontal.

Example

Charley's amazing floating home is travelling at a speed of 5 ms⁻¹ at an angle of 60° up from the horizontal. Find the vertical and horizontal components.

Charley's mobile home was the envy of all his friends.

The **horizontal** component v_x, is:

$v_x = v \cos \theta = \mathbf{5 \cos 60°} = 2.5$ ms⁻¹

The vertical component v_y is:

$v_y = v \sin \theta = \mathbf{5 \sin 60°} = 4.33$ ms⁻¹

Resolving is dead useful because the two components of a vector **don't affect each other**. This means you can deal with the two directions **completely separately**.

Only the vertical component is affected by gravity.

Practice Questions

Q1   Explain the difference between a scalar quantity and a vector quantity.

Q2   Jemima has gone for a swim in a river which is flowing at 0.35 ms⁻¹. She swims at 0.18 ms⁻¹ at right angles to the current. Show that her resultant velocity is 0.39 ms⁻¹ at an angle of 27.2° to the current.

Q3   Jemima is pulling on her lead with a force of 40 N at an angle of 26° below the horizontal. Show that the horizontal component of this force is about 36 N.

Exam Questions

Q1   The wind is creating a horizontal force of 20 N on a falling rock of weight 75 N. Calculate the magnitude and direction of the resultant force.   [2 marks]

Q2   A glider is travelling at a velocity of 20.0 ms⁻¹ at an angle of 15° below the horizontal. Find the horizontal and vertical components of the glider's velocity.   [2 marks]

His Dark Vectors Trilogy — displacement, velocity and acceleration...

Well there's nothing like starting the book on a high. And this is nothing like... yes, OK. Ahem. Well, good evening folks. I'll mostly be handing out useful information in boxes like this. But I thought I'd not rush into it, so this one's totally useless.

Motion with Constant Acceleration

Uniform Acceleration is Constant Acceleration

Acceleration could mean a change in speed or direction or both.

Uniform means **constant** here. It's nothing to do with what you wear.

There are **four main equations** that you use to solve problems involving **uniform acceleration**. You need to be able to use them, **and** how they're **derived**.

1) **Acceleration is the rate of change of velocity.**

From this definition you get:

$$a = \frac{(v - u)}{t} \quad \text{so} \quad \boxed{v = u + at}$$

where:

u = initial velocity a = acceleration
v = final velocity t = time taken

2) **s = average velocity × time**

If acceleration is constant, the average velocity is just the average of the initial and final velocities, so:

$$\boxed{s = \frac{(u + v)}{2} \times t} \quad s = \text{displacement}$$

3) Substitute the expression for v from equation 1 into equation 2 to give:

$$s = \frac{(u + u + at) \times t}{2} = \frac{2ut + at^2}{2} \qquad \boxed{s = ut + \tfrac{1}{2}at^2}$$

4) You can **derive** the fourth equation from equations **1** and **2**:

Use equation **1** in the form:

$$a = \frac{v - u}{t}$$

Multiply both sides by s, where:

$$s = \frac{(u + v)}{2} \times t$$

This gives us:

$$as = \frac{(v - u)}{t} \times \frac{(u + v)t}{2}$$

The t's on the right cancel, so:

$$2as = (v - u)(v + u)$$
$$2as = v^2 - uv + uv - u^2$$

so: $\boxed{v^2 = u^2 + 2as}$

Example

A tile falls from a roof 25 m high. Calculate its speed when it hits the ground and how long it takes to fall. Take **g** = 9.8 ms^{-2}.

First of all, write out what you know:

s = 25 m

u = 0 ms^{-1} since the tile's stationary to start with

a = 9.8 ms^{-2} due to gravity

v = ? t = ?

Usually you take upwards as the positive direction. In this question it's probably easier to take downwards as positive, so you get g = +9.8 ms^{-2} instead of g = −9.8 ms^{-2}.

9.8 ms^{-2}

25 m

Then, choose an equation with only **one unknown quantity**.

So start with $v^2 = u^2 + 2as$

$v^2 = 0 + 2 \times 9.8 \times 25$

$v^2 = 490$

$v = 22.1$ ms^{-1}

Now, find t using:

$s = ut + \tfrac{1}{2}at^2$

$25 = 0 + \tfrac{1}{2} \times 9.8 \times t^2$

$t^2 = \dfrac{25}{4.9}$

Final answers:

$t = 2.3$ s

$v = 22.1$ ms^{-1}

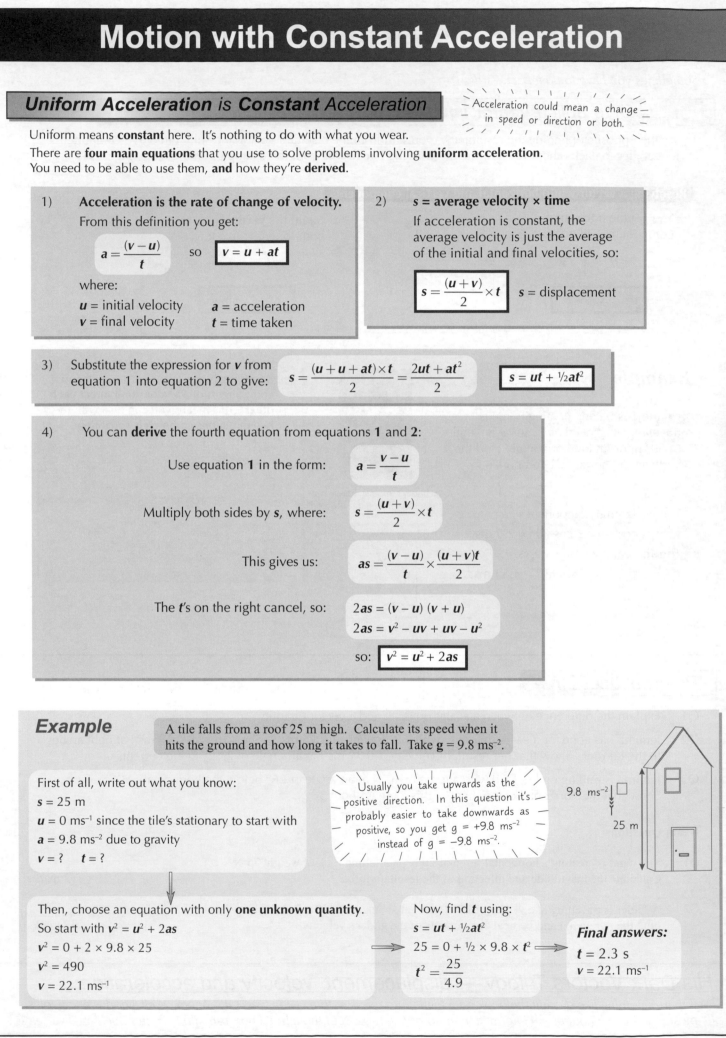

Motion with Constant Acceleration

Example

A car accelerates steadily from rest at a rate of 4.2 ms^{-2} for 6 seconds.

a) Calculate the final speed.

b) Calculate the distance travelled in 6 seconds.

4.2 ms^{-2}

Remember — always start by writing down what you know.

a) $a = 4.2$ ms^{-2} choose the right equation... $v = u + at$

$u = 0$ ms^{-1} $v = 0 + 4.2 \times 6$

$t = 6$ s   **Final answer:** $v = 25.2$ ms^{-1}

$v = ?$

b) $s = ?$ you can use: $s = \dfrac{(u+v)t}{2}$ or: $s = ut + \frac{1}{2}at^2$

$t = 6$ s

$u = 0$ ms^{-1}

$a = 4.2$ ms^{-2} $s = \dfrac{(0+25.2) \times 6}{2}$ $s = 0 + \frac{1}{2} \times 4.2 \times (6)^2$

$v = 25.2$ ms^{-1}

Final answer: $s = 75.6$ m $s = 75.6$ m

You Have to **Learn** the Constant Acceleration **Equations**

Make sure you learn the equations. There are only four of them and these questions are always dead easy marks in the exam, so you'd be dafter than a hedgehog in a helicopter not to learn them...

Practice Questions

Q1 Write out the four constant acceleration equations.

Q2 Show how the equation $s = ut + \frac{1}{2}at^2$ can be derived.

Exam Questions

Q1 A skydiver jumps from an aeroplane when it is flying horizontally. She accelerates due to gravity for 5 s.
(a) Calculate her maximum vertical velocity. (Assume no air resistance.)   [2 marks]
(b) How far does she fall in this time?   [2 marks]

Q2 A motorcyclist slows down uniformly as he approaches a red light. He takes 3.2 seconds to come to a halt and travels 40 m in this time.
(a) How fast was he travelling initially?   [2 marks]
(b) Calculate his acceleration. (N.B. a negative value shows a deceleration.)   [2 marks]

Q3 A stream provides a constant acceleration of 6 ms^{-2}. A toy boat is pushed directly against the current from a point 1.2 m upstream from a small waterfall, then released. Just before it reaches the waterfall, it is travelling at a speed of 5 ms^{-1}.
(a) Find the initial velocity of the boat.   [2 marks]
(b) What is the maximum distance upstream from the waterfall the boat reaches?   [2 marks]

Constant acceleration — it'll end in tears...

If a question talks about "uniform" or "constant" acceleration, it's a dead giveaway they want you to use one of these equations. The tricky bit is working out which one to use — start every question by writing out what you know and what you need to know. That makes it much easier to see which equation you need. To be sure. Arrr.

Free Fall

So, how do you work this parachute thing agaiAAAAAaaaaaaarrrrrrgggghhhhhhhhhhhhhhh...

Free Fall is when there's Only Gravity and Nothing Else

Free fall is defined as "the motion of an object undergoing an acceleration of '*g*'".
You need to remember:

1) Acceleration is a **vector quantity** — and '*g*' acts **vertically downwards**.

2) Unless you're given a different value, take the magnitude of *g* as **9.81 ms⁻²**, though it varies slightly at different points on the Earth's surface.

3) The **only force** acting on an object in free fall is its **weight**.

4) Objects can have an initial velocity in any direction and still undergo **free fall** as long as the **force** providing the initial velocity is **no longer acting**.

You Can Measure g by using an Object in Free Fall

You don't have to do it this way — but if you don't know a method of measuring *g* already, learn this one.

You need to be able to:

1) **Sketch** a diagram of the **apparatus**.

2) **Describe** the **method**.

3) **List** the **measurements** you make.

4) **Explain** how '*g*' is **calculated**.

5) Be aware of sources of **error**.

Another gravity experiment.

Experiment to Measure the Acceleration Due to Gravity

electromagnet

ball bearing

switch

height *h*

timer

trapdoor

In this experiment you have to assume that the effect of air resistance on the ball bearing is negligible and that the magnetism of the electromagnet decays instantly.

The Method:

1) Measure the height *h* from the **bottom** of the ball bearing to the **trapdoor**.

2) Flick the switch to simultaneously start the timer and disconnect the electromagnet, releasing the ball bearing.

3) The ball bearing falls, knocking the trapdoor down and breaking the circuit — which stops the timer.

Use the time *t* measured by the timer, and the height *h* that the ball bearing has fallen, to calculate a value for **g**, using $h = \frac{1}{2}gt^2$

(see next page for more on acceleration formulas).

The most significant source of error in this experiment will be in the measurement of *h*.
Using a ruler, you'll have an uncertainty of about 1 mm.
This dwarfs any error from switch delay or air resistance.

Free Fall

You can Just **Replace a** with **g** in the **Equations of Motion**

You need to be able to work out **speeds**, **distances** and **times** for objects in **free fall**. Since *g* is a **constant acceleration** you can use the **constant acceleration equations**. But *g* acts downwards, so you need to be careful about directions.

To make it clear, there's a sign convention: **upwards is positive**, **downwards is negative**.

> **Sign Conventions — Learn Them:**
>
> *g* is always <u>downwards</u> so it's <u>usually negative</u>   *t* is <u>always positive</u>
>
> *u* and *v* can be either <u>positive or negative</u>   *s* can be either <u>positive or negative</u>

Case 1: No initial velocity (it just falls)

Initial velocity *u* = 0

Acceleration $a = g = -9.81 \text{ ms}^{-2}$

So the constant acceleration equations become: $\Longrightarrow$

$$v = gt \qquad v^2 = 2gs$$
$$s = \frac{1}{2}gt^2 \qquad s = \frac{vt}{2}$$

Case 2: An initial velocity upwards (it's thrown up into the air)

The constant acceleration equations are just as normal, but with $a = g = -9.81 \text{ ms}^{-2}$

Case 3: An initial velocity downwards (it's thrown down)

Example: Alex throws a stone down a cliff. She gives it a downwards velocity of 2 ms⁻¹. It takes 3 s to reach the water below. How high is the cliff?

1) You know $u = -2 \text{ ms}^{-1}$, $a = g = -9.81 \text{ ms}^{-2}$ and $t = 3$ s. You need to find **s**. ← *s* will be negative because the stone ends up further down than it started

2) Use $s = ut + \frac{1}{2}gt^2 = (-2 \times 3) + \left(\frac{1}{2} \times -9.81 \times 3^2\right) = -50.1$ m. **The cliff is 50.1 m high.**

Practice Questions

Q1   What is the value of the acceleration of a free-falling object?

Q2   What is the initial velocity of an object which is dropped?

Exam Questions

Q1   A student has designed a device to estimate the value of '*g*'. It consists of two narrow strips of card joined by a piece of transparent plastic. The student drops the device through a light gate connected to a computer. As the device falls, the strips of card break the light beam.

(a) What three pieces of data will the student need from the computer to estimate '*g*'?   [3 marks]

(b) Explain how the measurements from the light gate can be used to estimate '*g*'.   [3 marks]

(c) Give one reason why the student's value of '*g*' will not be entirely accurate.   [1 mark]

Q2   Charlene is bouncing on a trampoline. She reaches her highest point a height of 5 m above the trampoline.

(a) Calculate the speed with which she leaves the trampoline surface.   [2 marks]

(b) How long does it take her to reach the highest point?   [2 marks]

(c) What will her velocity be as she lands back on the trampoline?   [2 marks]

It's not the falling that hurts — *it's the* being pelted with rotten vegetables... okay, okay...

The hardest bit with free fall questions is getting your signs right. Draw yourself a little diagram before you start doing any calculations, just label it with what you know and what you want to know. That can help you get the signs straight in your head. It also helps the person marking your paper if it's clear what your sign convention is. Always good.

Free Fall and Projectile Motion

What goes up, must come down — but no one really questioned why until Aristotle. He thought he knew... but then Galileo and Newton sure showed him...

Aristotle — *Heavy* Objects Fall *Quicker* than *Lighter* objects

1) **Aristotle** was an ancient Greek philosopher who sat around thinking about pretty much everything, including the **joys of Physics**.

2) He used **reasoning** to try and work out how the world worked from **everyday** observations.

3) One of his famous theories was that if **two objects** of **different mass** are dropped from the **same height**, the **heavier** object would always hit the ground **before** the lighter object.

Trev was counting... there was no way Tez took longer to hit the ground than him.

Galileo — *All* Objects in Free Fall Accelerate *Uniformly*

1) Galileo thought that **all objects accelerate towards the ground at the same rate** — so objects with different weights dropped from the same height should hit the ground at the **same time**.

2) Not only that, but he reckoned the reason objects didn't seem to do this was because of the effect of **air resistance** on different objects.

3) Believe it or not, scientists don't think Galileo chucked stuff from the top of the Leaning Tower of Pisa to test this theory. Instead he did an even more exciting experiment — he.. er... rolled balls down a slope.

Example: The Inclined Plane Experiment

1) Handy things like stop clocks and light gates hadn't been invented, so Galileo had to find a way of slowing down the free fall of an object without otherwise affecting to have any chance of showing it accelerating.

2) Ta da — the inclined plane experiment was born. Galileo found that by rolling a ball down a **smooth** groove in an inclined plane, he **reduced** the effect of **air resistance** while slowing the ball's fall at the same time.

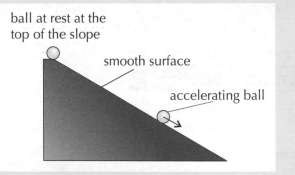

ball at rest at the top of the slope

smooth surface

accelerating ball

3) He timed how long it took the ball to roll from the top of the slope to the bottom using a water clock.

4) By rolling the ball along different fractions of the total length of the slope, he found that the distance the ball travelled was proportional to the square of the time taken. The ball was **accelerating** at a **constant rate**.

5) In the end it took Newton to bring it all together and explain why **all** free-falling objects have the **same acceleration** (see page 17).

Galileo *Tested* his *Theories* using *Experiments*

1) Not only did Galileo disagree with Aristotle on almost everything, he liked to shout it from the rooftops. He not only managed to insult other philosophers at the time, but the Pope and the entire Catholic church too — which got him in a whole **load of trouble**.

2) Even though he was so unpopular, Galileo's theories eventually overturned Aristotle's and became **generally accepted**. He wasn't the first person to question Aristotle, but his success was down to the **systematic** and **rigorous experiments** he used to **test** his theories. These experiments could be repeated and the results described **mathematically** and compared.

3) It's not much different in science now. You make a prediction and test it — the more **scientific evidence** that supports your theory, the more **accepted** it becomes.

Free Fall and Projectile Motion

Any object given an initial velocity and then left to move freely under gravity is a projectile.
If you're doing AS Maths, you've got all this to look forward to in M1 as well, quite likely. Fun for all the family.

You have to think of **Horizontal** and **Vertical** Motion **Separately**

Example

Sharon fires a scale model of a TV talent show presenter horizontally with a velocity of $100\,\text{ms}^{-1}$ from 1.5 m above the ground. How long does it take to hit the ground, and how far does it travel? Assume the model acts as a particle, the ground is horizontal and there is no air resistance.

Think about vertical motion first:

1) It's **constant acceleration** under gravity...

2) You know $u = 0$ (no vertical velocity at first), $s = -1.5$ m and $a = g = -9.81\,\text{ms}^{-2}$. You need to find t.

3) Use $s = \dfrac{1}{2}gt^2 \Rightarrow t = \sqrt{\dfrac{2s}{g}} = \sqrt{\dfrac{2 \times -1.5}{-9.81}} = 0.55$ s

4) So the model hits the ground after **0.55** seconds.

Then do the horizontal motion:

1) The horizontal motion isn't affected by gravity or any other force, so it moves at a **constant speed**.

2) That means you can just use good old **speed = distance / time**.

3) Now $v_h = 100\,\text{ms}^{-1}$, $t = 0.55$ s and $a = 0$. You need to find s_h.

4) $s_h = v_h t = 100 \times 0.55 = \underline{55\text{ m}}$

Where v_h is the horizontal velocity, and s_h is the horizontal distance travelled (rather than the height fallen).

It's **Slightly Trickier** if it **Starts Off** at an **Angle**

If something's projected at an angle (like, say, a javelin) you start off with both horizontal and vertical velocity:

Method:
1) Resolve the initial velocity into horizontal and vertical components.
2) Use the vertical component to work out how long it's in the air and/or how high it goes.
3) Use the horizontal component to work out how far it goes while it's in the air.

Practice Questions

Q1 What is the initial vertical velocity for an object projected horizontally with a velocity of 5 ms^{-1}?

Q2 How does the horizontal velocity of a free-falling object change with time?

Q3 What is the main reason Galileo's ideas became generally accepted in place of Aristotle's?

Exam Questions

Q1 Jason stands on a vertical cliff edge throwing stones into the sea below.
He throws a stone horizontally with a velocity of 20 ms^{-1}, 560 m above sea level.
(a) How long does it take for the stone to hit the water from leaving Jason's hand?
Use g = 9.81 ms^{-2} and ignore air resistance. [2 marks]
(b) Find the distance of the stone from the base of the cliff when it hits the water. [2 marks]

Q2 Robin fires an arrow into the air with a vertical velocity of 30 ms^{-1}, and a horizontal velocity of 20 ms^{-1}, from 1 m above the ground. Find the maximum height from the ground reached by his arrow.
Use g = 9.81 ms^{-2} and ignore air resistance. [3 marks]

So it's this "Galileo" geezer who's to blame for my practicals...

Ah, the ups and downs and er... acrosses of life. Make sure you're happy splitting an object's motion into horizontal and vertical bits — it comes up all over mechanics. Hmmm... I wonder what Galileo would be proudest of, insisting on the systematic, rigorous experimental method on which modern science hangs... or getting in a Queen song? Magnificoooooo...

Displacement-Time Graphs

Drawing graphs by hand — oh joy. You'd think examiners had never heard of the graphical calculator.
Ah well, until they manage to drag themselves out of the dark ages, you'll just have to grit your teeth and get on with it.

Acceleration Means a Curved Displacement-Time Graph

A graph of displacement against time for an **accelerating object** always produces a **curve**.
If the object is accelerating at a **uniform rate**, then the **rate of change** of the **gradient** will be constant.

Example   Plot a displacement-time graph for a lion who accelerates constantly from rest at 2 ms⁻² for 5 seconds.

You want to find **s**, and you know that:
$a = 2$ ms⁻²
$u = 0$ ms⁻¹

Use $s = ut + \frac{1}{2}at^2$

If you substitute in **u** and **a**, this simplifies to:
$s = 0 \times t + \frac{1}{2} \times 2t^2$
$s = t^2$

Do a **table of values**:

t (s)	s (m)
0	**0**
1	**1**
2	**4**
3	**9**
4	**16**
5	**25**

...then plot the **graph**:

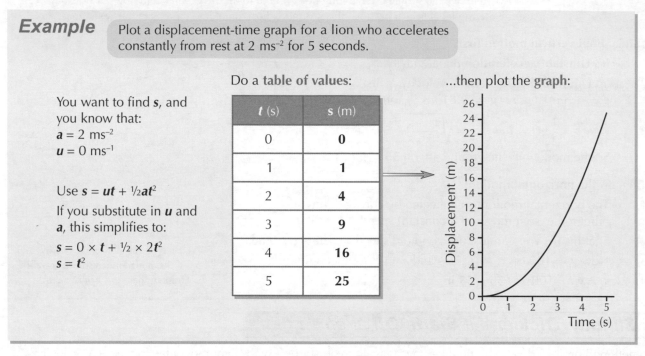

Different Accelerations Have Different Gradients

In the example above, if the lion has a **different acceleration** it'll change the **gradient** of the curve like this:

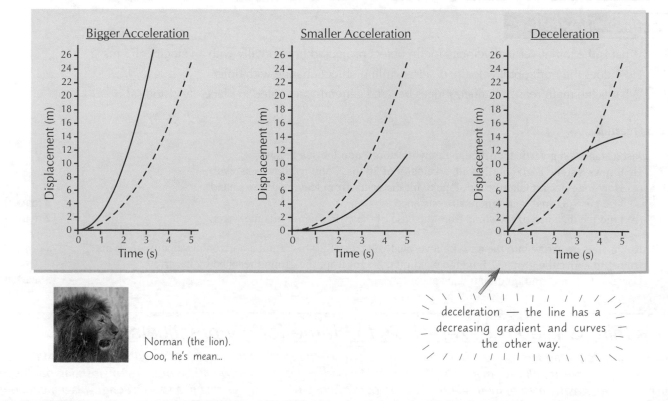

Bigger Acceleration   Smaller Acceleration   Deceleration

Norman (the lion).
Ooo, he's mean...

deceleration — the line has a decreasing gradient and curves the other way.

Displacement-Time Graphs

The *Gradient* of a *Displacement-Time Graph* Tells You the Velocity

When the velocity is constant, the graph's a **straight line**.
Velocity is defined as...

$$\text{velocity} = \frac{\text{change in displacement}}{\text{time taken}}$$

On the graph, this is $\frac{\text{change in } y\ (\Delta y)}{\text{change in } x\ (\Delta x)}$, i.e. the gradient.

So to get the velocity from a displacement-time graph,
just find the gradient.

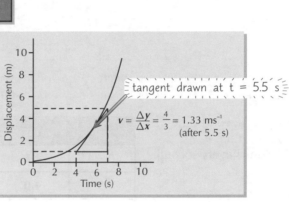

$$v = \frac{\Delta y}{\Delta x}$$
$$= \frac{10}{6} = 1.67 \text{ ms}^{-1}$$

It's the Same with *Curved Graphs*

If the gradient **isn't constant** (i.e. if it's a curved
line), it means the object is **accelerating**.

To find the **velocity** at a certain point you
need to draw a **tangent** to the curve at
that point and find its gradient.

tangent drawn at t = 5.5 s

$$v = \frac{\Delta y}{\Delta x} = \frac{4}{3} = 1.33 \text{ ms}^{-1}$$
(after 5.5 s)

Practice Questions

Q1   What is given by the slope of a displacement-time graph?

Q2   Sketch a displacement-time graph to show: a) constant velocity,   b) acceleration,   c) deceleration

Exam Questions

Q1   Describe the motion of the cyclist as shown by the graph below.   [4 marks]

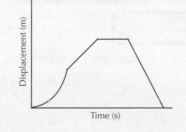

Q2   A baby crawls 5 m in 8 seconds at a constant velocity.   She then rests for 5 seconds before crawling a further
3 m in 5 seconds. Finally, she makes her way back to her starting point in 10 seconds, travelling at a constant
speed all the way.
(a) Draw a displacement-time graph to show the baby's journey.   [4 marks]
(b) Calculate her velocity at all the different stages of her journey.   [2 marks]

Some curves are bigger than others...

*Whether it's a straight line or a curve, the steeper it is, the greater the velocity. There's nothing difficult about these graphs
— the main problem is that it's easy to get them muddled up with velocity-time graphs (next page). If in doubt, think about
the gradient — is it velocity or acceleration, is it changing (curve), is it constant (straight line), is it 0 (horizontal line)...*

Velocity-Time Graphs

Speed-time graphs and velocity-time graphs are pretty similar.
The big difference is that velocity-time graphs can have a
negative part to show something travelling in the opposite direction:

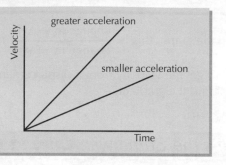

The **Gradient** of a **Velocity-Time Graph** tells you the **Acceleration**

$$\text{acceleration} = \frac{\text{change in velocity}}{\text{time taken}}$$

likewise for a speed-time graph

So the acceleration is just the **gradient** of a **velocity-time graph**.

1) **Uniform** acceleration is always a **straight line**.
2) The **steeper** the **gradient**, the **greater** the **acceleration**.

Example

A lion strolls along at 1.5 ms⁻¹ for 4 s and then accelerates uniformly at a
rate of 2.5 ms⁻² for 4 s. Plot this information on a velocity-time graph.

So, for the first four seconds, the
velocity is 1.5 ms⁻¹, then it
increases by **2.5 ms⁻¹ every second**:

t (s)	v (ms⁻¹)
0 – 4	1.5
5	4.0
6	6.5
7	9.0
8	11.5

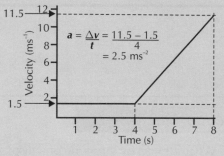

Derek (the lion)...

$$a = \frac{\Delta v}{t} = \frac{11.5 - 1.5}{4}$$
$$= 2.5 \text{ ms}^{-2}$$

You can see that the **gradient of the line** is **constant** between 4 s and 8 s and has a value of 2.5 ms⁻²,
representing the **acceleration of the lion**.

Distance *Travelled* = **Area** *under* **Speed-Time Graph**

You know that:

$$\text{distance travelled} = \text{average speed} \times \text{time}$$

So you can find the distance travelled by working out the **area under a speed-time graph**.

Example

A racing car accelerates uniformly from rest to 40 ms⁻¹ in 10 s. It maintains this speed for a further
20 s before coming to rest by decelerating at a constant rate over the next 15 s. Draw a velocity-time
graph for this journey and use it to calculate the total distance travelled by the racing car.

Split the **graph** up into **sections**: A, B and C
Calculate the **area** of each and **add** the three results together.
A: Area = ½ base × height = ½ × 10 × 40 = 200 m
B: Area = b × h = 20 × 40 = 800 m
C: Area = ½ b × h = ½ × 15 × 40 = 300 m
Total distance travelled = 1300 m

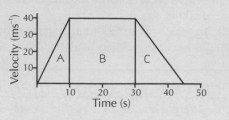

Velocity-Time Graphs

Non-Uniform Acceleration is a Curve on a V-T Graph

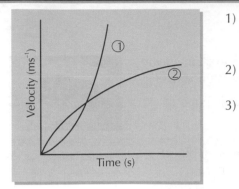

1) If the acceleration is changing, the gradient of the velocity-time graph will also be changing — so you **won't** get a **straight line**.

2) **Increasing acceleration** is shown by an **increasing gradient** — like in curve ①.

3) **Decreasing acceleration** is shown by a **decreasing gradient** — like in curve ②.

Simple enough...

You Can Draw Displacement-Time and Velocity-Time Graphs Using ICT

Instead of gathering distance and time data using **traditional methods**, e.g. a stopwatch and ruler, you can be a bit more **high-tech**.

A fairly **standard** piece of kit you can use for motion experiments is an **ultrasound position detector**. This is a type of **data-logger** that automatically records the **distance** of an object from the sensor several times a second.

If you attach one of these detectors to a computer with **graph-drawing software**, you can get **real-time** displacement-time and velocity-time graphs.

> The main **advantages** of data-loggers over traditional methods are:
>
> 1) The data is more **accurate** — you don't have to allow for human reaction times.
>
> 2) Automatic systems have a much higher **sampling** rate than humans — most ultrasound position detectors can take a reading ten times every second.
>
> 3) You can see the data displayed in **real time**.

Practice Questions

Q1 How do you calculate acceleration from a velocity-time graph?

Q2 How do you calculate the distance travelled from a speed-time graph?

Q3 Sketch velocity-time graphs for constant velocity and constant acceleration.

Q4 Describe the main advantages of ICT over traditional methods for the collection and display of motion data.

Exam Question

Q1 A skier accelerates uniformly from rest at 2 ms^{-2} down a straight slope.

(a) Sketch a velocity-time graph for the first 5 s of his journey. [2 marks]

(b) Use a constant acceleration equation to calculate his displacement at t = 1, 2, 3, 4 and 5 s, and plot this information onto a displacement-time graph. [5 marks]

(c) Suggest another method of calculating the skier's distance travelled after each second and use this to check your answers to part (b). [2 marks]

Still awake — I'll give you five more minutes...

There's a really nice sunset outside my window. It's one of those ones that makes the whole landscape go pinky-yellowish. And that's about as much interest as I can muster on this topic. Normal service will be resumed on page 16, I hope.

Newton's Laws of Motion

You did most of this at GCSE, but that doesn't mean you can just skip over it now. You'll be kicking yourself if you forget this stuff in the exam — easy marks...

Newton's Laws are Only Approximations

Isaac Newton's a pretty famous chap. Not only was he inspired by an apple to explain gravity, he also wrote three really important scientific laws — you need to know about the first two.

1) Newton's laws work pretty well. At **everyday speeds** they give really, really good approximations — but they're **not** the whole story.

2) At very **high speeds**, you have to take into account **relativistic effects**.
According to the **Special Theory of Relativity**, as you increase the speed of an object its **mass** increases. So mass isn't constant, and $F = ma$ doesn't work any more.

Gerty swore she lost
10 pounds as soon as she
stopped running.

Newton's 1st Law says that a Force is Needed to Change Velocity

1) **Newton's 1st law of motion** states that the **velocity** of an object will **not change** unless a **net force** acts on it.

2) In plain English this means a body will stay still or move in a **straight line** at a **constant speed**, unless there's a **net force** acting on it.

3) If the forces **aren't balanced**, the **overall net force** will make the body **accelerate**. This could be a change in **direction**, or **speed**, or both. (See Newton's 2nd law, below.)

An apple sitting on a table won't go anywhere because the **forces** on it are **balanced**.

reaction (R)	=	**weight** (mg)
(force of table pushing apple up)		(force of gravity pulling apple down)

Newton's 2nd Law says that Acceleration is Proportional to the Force

...which can be written as the well-known equation:

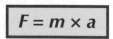

net force (N) = mass (kg) × acceleration (ms⁻²) $F = m \times a$

Learn this — it crops up all over the place in AS Physics. And learn what it means too:

1) It says that the **more force** you have acting on a certain mass, the **more acceleration** you get.

2) It says that for a given force the **more mass** you have, the **less acceleration** you get.

3) Galileo said that all objects fall at the same rate if you ignore air resistance — and that man wasn't wrong. You can see **why** with a bit of ball dropping and a dash of Newton's 2nd law — check out the next page.

> *REMEMBER:*
> 1) The **net force** is the **vector sum** of all the forces.
> 2) The force is **always** measured in **newtons**.
> 3) The **mass** is always measured in **kilograms**.
> 4) The **acceleration** is always in the **same direction** as the **net force**.

Newton's Laws of Motion

Acceleration *is* Independent *of Mass*

Imagine dropping two balls at the same time — ball **1** being heavy, and ball **2** being light. Then use Newton's 2nd law to find their acceleration.

mass = m_1 resultant force = F_1 acceleration = a_1 By Newton's second law: $\qquad F_1 = m_1a_1$ Ignoring air resistance, the only force acting on the ball is weight, given by $W_1 = m_1g$ (where g = gravitational field strength = 9.81 Nkg^{-1}). So: $F_1 = m_1a_1 = W_1 = m_1g$ So: $m_1a_1 = m_1g$, then m_1 cancels out to give: $a_1 = g$	mass = m_2 resultant force = F_2 acceleration = a_2 By Newton's second law: $\qquad F_2 = m_2a_2$ Ignoring air resistance, the only force acting on the ball is weight, given by $W_2 = m_2g$ (where g = gravitational field strength = 9.81 Nkg^{-1}). So: $F_2 = m_2a_2 = W_2 = m_2g$ So: $m_2a_2 = m_2g$, then m_2 cancels out to give: $a_2 = g$

... in other words, the **acceleration** is **independent of the mass**. It makes **no difference** whether the ball is **heavy or light**. And I've kindly **hammered home the point** by showing you two almost identical examples.

Practice Questions

Q1   State Newton's 1st and 2nd laws of motion, and explain what they mean.

Q2   Ball A has a mass of 5 kg and ball B has a mass of 3 kg.
Both balls are dropped from the same height at the same time — which one hits the ground first? Why?

Exam Questions

Q1   Draw diagrams to show the forces acting on a parachutist:
    (i)   accelerating downwards.   [1 mark]
    (ii) having reached terminal velocity.   [1 mark]

Q2   A boat is moving across a river. The engines provide a force of 500 N at right angles to the flow of the river and the boat experiences a drag of 100 N in the opposite direction. The force on the boat due to the flow of the river is 300 N. The mass of the boat is 250 kg.
    (a) Calculate the magnitude of the net force acting on the boat.   [2 marks]
    (b) Calculate the magnitude of the acceleration of the boat.   [2 marks]

Q3   This question asks you to use Newton's second law to explain three situations.
    (a) Two cars have different maximum accelerations.
        What are the only two overall factors that determine the acceleration a car can have?   [2 marks]
    (b) Michael can always beat his younger brother Tom in a sprint, however short the distance.
        Give two possible reasons for this.   [2 marks]
    (c) Michael and Tom are both keen on diving. They notice that they seem to take the same time to drop
        from the diving board to the water. Explain why this is the case.   [3 marks]

Newton's 4th Law... avoid the brain-eating apples falling from the sky...

These laws may not really fill you with a huge amount of excitement (and I hardly blame you if they don't)... but it was pretty fantastic at the time — suddenly people actually understood how forces work, and how they affect motion. I mean arguably it was one of the most important scientific discoveries ever...

Drag and Terminal Velocity

If you jump out of a plane at 2000 ft, you want to know that you're not going to be accelerating all the way.

Friction is a Force that Opposes Motion

There are two main types of friction:

1) **Contact friction** between **solid surfaces** (which is what we usually mean when we just use the word 'friction'). You don't need to worry about that too much for now.

2) **Fluid friction** (known as **drag** or fluid resistance or air resistance).

> **Fluid Friction or Drag:**
>
> 1) 'Fluid' is a word that means either a **liquid or a gas** — something that can **flow**.
> 2) The force depends on the thickness (or **viscosity**) of the fluid.
> 3) It **increases** as the **speed increases** (for simple situations it's directly proportional, but you don't need to worry about the mathematical relationship).
> 4) It also depends on the **shape** of the object moving through it — the larger the **area** pushing against the fluid, the greater the resistance force.

Things you need to remember about frictional forces:

1) They **always** act in the **opposite direction** to the **motion** of the object.
2) They can **never** speed things up or start something moving.
3) They convert **kinetic energy** into **heat**.

Terminal Velocity — when the Friction Force Equals the Driving Force

You will reach a **terminal velocity** at some point, if you have:

1) a **driving force** that stays the **same** all the time
2) a **frictional** or **drag force** (or collection of forces) that increases with speed

There are **three main stages** to reaching terminal velocity:

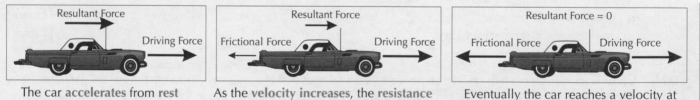

The car **accelerates** from **rest** using a constant driving force.	As the **velocity increases**, the **resistance forces increase** (because of things like turbulence — you don't need the details). This **reduces the resultant force** on the car and hence **reduces its acceleration**.	Eventually the car reaches a velocity at which the **resistance forces are equal to the driving force**. There is now **no resultant force** and **no acceleration**, so the car carries on at **constant velocity**.

Sketching a Graph for Terminal Velocity

You need to be able to **recognise** and **sketch** the graphs for **velocity against time** and **acceleration against time** for the **terminal velocity** situation.

Nothing for it but practice — shut the book and sketch them from memory. Keep doing it till you get them right every time.

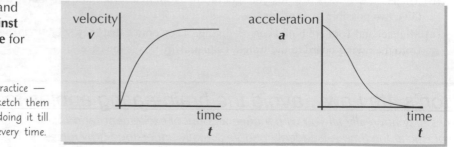

Drag and Terminal Velocity

Things **Falling** through **Air** or **Water** Reach a **Terminal Velocity** too

When something's falling through **air**, the **weight** of the object is a **constant force** accelerating the object downwards.
Air resistance is a **frictional force** opposing this motion, which **increases** with **speed**.
So before a parachutist opens the parachute, exactly the same thing happens as with the car example:

1) A skydiver leaves a plane and will **accelerate** until the **air resistance** equals his **weight**.

driving force
air resistance

2) He will then be travelling at a **terminal velocity**.

driving force
air resistance

But... the terminal velocity of a person in free fall is too great to land without dying a horrible death.
The **parachute increases** the **air resistance massively**, which slows him down to a lower terminal velocity:

3) Before reaching the ground he will **open his parachute**, which immediately **increases the air resistance** so it is now **bigger** than his **weight**.

driving force
air resistance

4) This **slows him down** until his speed has dropped enough for the **air resistance** to be **equal to his weight** again. This new terminal velocity is small enough to survive landing.

driving force
air resistance

The v-t graph is a bit different, because you have a new terminal velocity being reached after the parachute is opened:

velocity
v
Terminal velocity before parachute opened
Parachute opened
Terminal velocity after parachute opened
time
t

Practice Questions

Q1 What forces limit the speed of a skier going down a slope?

Q2 What conditions cause a terminal velocity to be reached?

Q3 Sketch a graph to show how the velocity changes with time for an object falling through air.

Exam Question

Q1 A space probe free-falls towards the surface of a planet.
The graph on the right shows the velocity data recorded by the probe as it falls.

velocity, **v**
time, **t**

(a) The planet does not have an atmosphere. Explain how you can tell this from the graph. [2 marks]

(b) Sketch the velocity-time graph you would expect to see if the planet did have an atmosphere. [2 marks]

(c) Explain the shape of the graph you have drawn. [3 marks]

You'll never understand this without going parachuting*...

*When you're doing questions about terminal velocity, remember the frictional forces reduce acceleration, not speed.
They usually don't slow an object down, apart from in the parachute example, where the skydiver is travelling faster
when the parachute opens than the terminal velocity for the parachute-skydiver combination.*

Mass, Weight and Centre of Gravity

I'm sure you know all this 'mass', 'weight' and 'density' stuff from GCSE. But let's just make sure...

The Mass of a Body makes it Resist Changes in Motion

1) The **mass** of an object is the **amount of 'stuff'** (or **matter**) in it. It's measured in **kg**.

2) The greater an object's mass, the greater its **resistance** to a **change in velocity** (called its **inertia**).

3) The **mass** of an object **doesn't change** if the strength of the **gravitational field** changes.

4) Weight is a **force**. It's measured in **newtons** (N), like all forces.

5) Weight is the **force experienced by a mass** due to a **gravitational field**.

6) The weight of an object **does vary** according to the size of the **gravitational field** acting on it.

> **weight = mass × gravitational field strength (W = mg)**   where g = 9.81 Nkg^{-1} on Earth.

This table shows Gerald (the lion*)'s
mass and weight on the Earth and the Moon.

Name	Quantity	Earth (g = 9.81 Nkg^{-1})	Moon (g = 1.6 Nkg^{-1})
Mass	Mass (scalar)	150 kg	150 kg
Weight	Force (vector)	1471.5 N	240 N

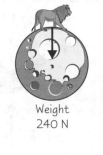

Weight
240 N

Weight
1470 N

Density is Mass per Unit Volume

Density is a measure of the 'compactness' (for want of a better word) of a substance.
It relates the mass of a substance to how much space it takes up.

> $$density = \frac{mass}{volume} \qquad \rho = \frac{m}{V}$$

The symbol for density is a Greek letter rho (ρ) — it looks like a p but it isn't.

The **units** of **density** are **$g\,cm^{-3}$** or **$kg\,m^{-3}$**
N.B. 1 $g\,cm^{-3}$ = 1000 $kg\,m^{-3}$

1) The density of an object depends on what it's made of.
Density **doesn't vary** with **size or shape**.

2) The **average density** of an object determines whether it **floats** or **sinks**.

3) A solid object will **float** on a fluid if it has a **lower density** than the **fluid**.

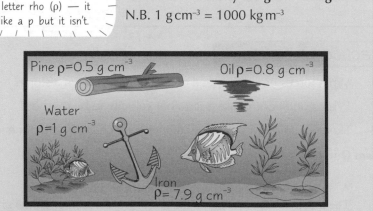

Pine ρ=0.5 g cm^{-3}

Oil ρ=0.8 g cm^{-3}

Water ρ=1 g cm^{-3}

Iron ρ= 7.9 g cm^{-3}

Centre of Gravity — Assume All the Mass is in One Place

1) The **centre of gravity** (or centre of mass) of an object is the **single point** that you can consider its **whole weight** to **act through** (whatever its orientation).

2) The object will always **balance** around this **point**, although in some cases the **centre of gravity** will **fall outside** the object.

Centre of gravity

Centre of gravity

Centre of gravity

*Yes, I know — I just like lions, OK...

Mass, Weight and Centre of Gravity

Find the Centre of Gravity either by Symmetry or Experiment

Experiment to find the Centre of Gravity of an Irregular Object

1) Hang the object freely from a point (e.g. one corner).
2) Draw a vertical line downwards from the point of suspension — use a plumb bob to get your line exactly vertical.
3) Hang the object from a different point.
4) Draw another vertical line down.
5) The centre of gravity is where the two lines cross.

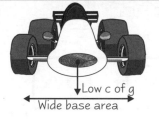

For a regular object you can just use symmetry. The centre of gravity of any regular shape is at its centre.

How High the Centre of Gravity is tells you How Stable the Object is

1) An object will be nice and **stable** if it has a **low centre** of **gravity** and a **wide base area**. This idea is used a lot in design, e.g. Formula 1 racing cars.

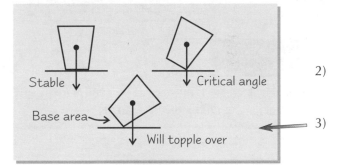

2) The **higher** the **centre of gravity**, and the **smaller** the **base area**, the **less stable** the object will be. Think of unicyclists...

3) An object will topple over if a **vertical line** drawn **downwards** from its **centre of gravity** falls **outside** its **base area**.

Practice Questions

Q1 A lioness has a mass of 200 kg. What would be her mass and weight on the Earth (where g = 9.8 Nkg⁻¹) and on the Moon (where g = 1.6 Nkg⁻¹)?

Q2 What is meant by the centre of gravity of an object?

Exam Questions

Q1 (a) Define **density**. [1 mark]

(b) A cylinder of aluminium, radius 4 cm and height 6 cm, has a mass of 820 g. Calculate its density. [3 marks]

(c) Use the information from part (b) to calculate the mass of a cube of aluminium of side 5 cm. [1 mark]

Q2 Describe an experiment to find the centre of gravity of an object of uniform density with a constant thickness and irregular cross-section. Identify one major source of uncertainty and suggest a way to reduce its effect on the accuracy of your result. [5 marks]

The centre of gravity of this book should be round about page 90...

This is a really useful area of physics. To would-be nuclear physicists it might seem a little dull, but if you want to be an engineer — something a bit more useful (no offence Einstein) — then things like centre of gravity and density are dead important things to understand. You know, for designing things like cars and submarines... yep, pretty useful I'd say.

Forces and Equilibrium

Remember the vector stuff from the beginning of the section... good, you're going to need it...

Resolving a Force means Splitting it into Components

1) **Forces** are **vector quantities** and so when you draw the forces on an object, the **arrow labels** should show the **size** and **direction** of the forces.

2) Forces can be in **any direction**, so they're not always at right angles to each other. This is sometimes a bit **awkward** for **calculations**.

3) To make an 'awkward' force easier to deal with, you can think of it as **two separate forces**, acting at **right angles** to **each other**.

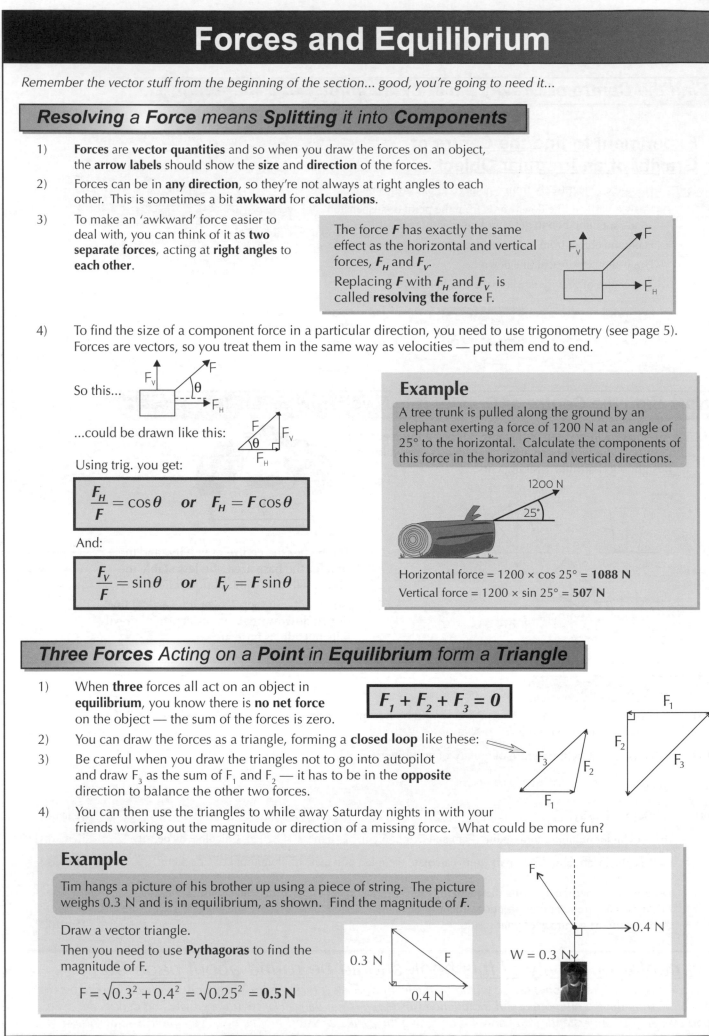

> The force **F** has exactly the same effect as the horizontal and vertical forces, **F_H** and **F_V**.
>
> Replacing **F** with **F_H** and **F_V** is called **resolving the force** F.

4) To find the size of a component force in a particular direction, you need to use trigonometry (see page 5). Forces are vectors, so you treat them in the same way as velocities — put them end to end.

So this...

...could be drawn like this:

Using trig. you get:

$$\frac{F_H}{F} = \cos\theta \quad \textbf{or} \quad F_H = F\cos\theta$$

And:

$$\frac{F_V}{F} = \sin\theta \quad \textbf{or} \quad F_V = F\sin\theta$$

Example

A tree trunk is pulled along the ground by an elephant exerting a force of 1200 N at an angle of 25° to the horizontal. Calculate the components of this force in the horizontal and vertical directions.

1200 N

25°

Horizontal force = 1200 × cos 25° = **1088 N**

Vertical force = 1200 × sin 25° = **507 N**

Three Forces Acting on a Point in Equilibrium form a Triangle

1) When **three** forces all act on an object in **equilibrium**, you know there is **no net force** on the object — the sum of the forces is zero.

$$F_1 + F_2 + F_3 = 0$$

2) You can draw the forces as a triangle, forming a **closed loop** like these:

3) Be careful when you draw the triangles not to go into autopilot and draw F_3 as the sum of F_1 and F_2 — it has to be in the **opposite** direction to balance the other two forces.

4) You can then use the triangles to while away Saturday nights in with your friends working out the magnitude or direction of a missing force. What could be more fun?

Example

Tim hangs a picture of his brother up using a piece of string. The picture weighs 0.3 N and is in equilibrium, as shown. Find the magnitude of **F**.

Draw a vector triangle.

Then you need to use **Pythagoras** to find the magnitude of F.

$$F = \sqrt{0.3^2 + 0.4^2} = \sqrt{0.25^2} = \textbf{0.5 N}$$

0.3 N   F

0.4 N

F

0.4 N

W = 0.3 N

Forces and Equilibrium

You *Add* the *Components Back Together* to get the *Resultant Force*

1) If **two forces** act on an object, you find the **resultant** (total) **force** by adding the **vectors** together and creating a **closed triangle**, with the resultant force represented by the **third side**.

2) Forces are vectors (as you know), so you use **vector addition** — draw the forces as vector arrows put 'tail to top'.

3) Then it's yet more trigonometry to find the **angle** and the **length** of the third side.

Example

Two dung beetles roll a dung ball along the ground at constant velocity.
Beetle A applies a force of 0.5 N northwards while beetle B
exerts a force of only 0.2 N eastwards.
What is the resultant force on the dung ball?

The resultant force is **0.54 N** at an angle of **21.8°** from North.

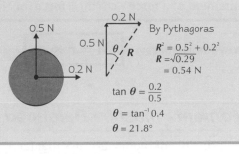

By Pythagoras
$R^2 = 0.5^2 + 0.2^2$
$R = \sqrt{0.29}$
$= 0.54$ N

$\tan \theta = \dfrac{0.2}{0.5}$

$\theta = \tan^{-1} 0.4$

$\theta = 21.8°$

Choose sensible *Axes* for *Resolving*

Use directions that **make sense** for the situation you're
dealing with. If you've got an object on a slope, choose
your directions **along the slope** and **at right angles to it**.
You can turn the paper to an angle if that helps.

Always choose sensible axes

Examiners like to call a
slope an "inclined plane".

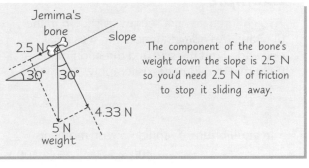

The component of the bone's
weight down the slope is 2.5 N
so you'd need 2.5 N of friction
to stop it sliding away.

Practice Questions

Q1 Sketch a free-body force diagram for an ice hockey puck moving
across the ice (assuming no friction).

Q2 What are the horizontal and vertical components of the force F?

Exam Questions

Q1 A picture is suspended from a hook as shown in the diagram.
Calculate the tension force, *T*, in the string.

[2 marks]

Q2 Two elephants pull a tree trunk as shown in the diagram.
Calculate the resultant force on the tree trunk.

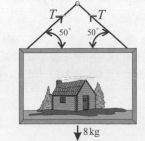

[2 marks]

Free-body force diagram — sounds like something you'd get with a dance mat...

Remember those F cos θ and F sin θ bits. Write them on bits of paper and stick them to your wall.
*Scrawl them on your pillow. Tattoo them on your brain. Whatever it takes — you just **have to learn them**.*

Moments and Torques

*This is not a time for jokes. There is not a moment to lose. The time for torquing is over. Oh ho ho ho ho *bang*. (Ow.)*

A **Moment** is the **Turning Effect** of a **Force**

The **moment**, or **torque**, of a **force** depends on the **size** of the force and **how far** the force is applied from the **turning point**:

> **moment of a force** (in Nm) = **force** (in N) × **perpendicular distance from pivot** (in m)

In symbols, that's:

$$M = F \times d$$

Moments *must be* Balanced *or the* Object *will* Turn

The **principle of moments** states that for a body to be in **equilibrium**, the **sum of the clockwise moments** about any point **equals** the **sum of the anticlockwise moments** about the same point.

Example

Two children sit on a seesaw as shown in the diagram. An adult balances the seesaw at one end. Find the size and direction of the force that the adult needs to apply.

1.5 m   1.0 m   0.5 m
400 N   300 N

In equilibrium, $\sum$ anticlockwise moments = $\sum$ clockwise moments

$\sum$ means "the sum of"

$$400 \times 1.5 = 300 \times 1 + 1.5F$$
$$600 = 300 + 1.5F$$

Final answer: $F = 200$ N downwards

Muscles, Bones *and* Joints *Act as* Levers

1) In a lever, an **effort force** (in this case from a muscle) acts against a **load force** (e.g. the weight of your arm) by means of a **rigid object** (the bone) rotating around a **pivot** (the joint).

2) You can use the **principle of moments** to answer lever questions:

Example

Find the force exerted by the biceps in holding a bag of gold still. The bag of gold weighs 100 N and the forearm weighs 20 N.

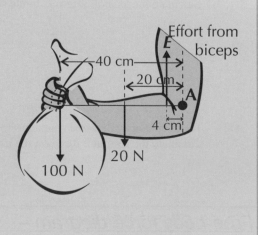

Effort from biceps
40 cm
20 cm
A
4 cm
100 N   20 N

Take moments about **A**.
In equilibrium:

$\sum$ anticlockwise moments = $\sum$ clockwise moments
$$(100 \times 0.4) + (20 \times 0.2) = 0.04E$$
$$40 + 4 = 0.04E$$

Final answer: $E = 1100$ N

Moments and Torques

A **Couple** is a **Pair** of **Forces**

1) A couple is a **pair** of **forces** of **equal size** which act **parallel** to each other, but in **opposite directions**.

2) A couple doesn't cause any resultant linear force, but **does** produce a **turning force** (usually called a **torque** rather than a moment).

The **size** of this **torque** depends on the **size** of the **forces** and the **distance** between them.

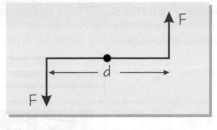

Torque of a couple (in Nm) = size of one of the forces (in N) × perpendicular distance between the forces (in m)

In symbols, that's: $T = F \times d$

Example

A cyclist turns a sharp right corner by applying equal but opposite forces of 20 N to the ends of the handlebars.

The length of the handlebars is 0.6 m.
Calculate the torque applied to the handlebars.

20 N

0.6 m

20 N

Torque = 20 × 0.6 = 12 Nm

Practice Questions

Q1 A girl of mass 40 kg sits 1.5 m from the middle of a seesaw.
Show that her brother, mass 50 kg, must sit 1.2 m from the middle if the seesaw is to balance.

Q2 What is meant by the word 'couple'?

Exam Questions

Q1 A driver is changing his flat tyre. The torque required to undo the nut is 60 Nm.
He uses a 0.4 m long double-ended wheel wrench.
Calculate the force that he must apply at each end of the spanner. [2 marks]

Q2 A diver of mass 60 kg stands on the end of a diving board 2 m from the pivot point.
Calculate the upward force exerted on the retaining spring 30 cm from the pivot.

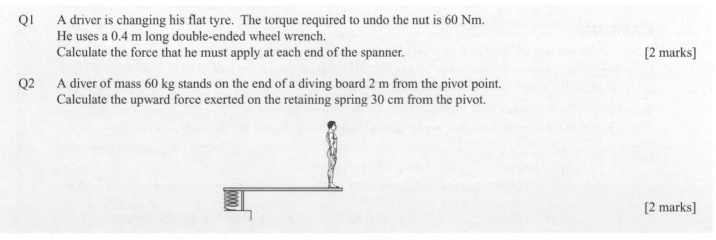

[2 marks]

It's all about balancing — just ask a tightrope walker...

They're always seesaw questions aren't they. It'd be nice if once, **just once**, they'd have a question on... I don't know, rotating knives or something. Just something unexpected... anything. It'd make physics a lot more fun, I'm sure. *sigh*

Car Safety

Some real applications now — how to avoid collisions, and how car manufacturers try to make sure you survive.

Many Factors Affect How Quickly a Car Stops

The braking distance and thinking distance together make the **total distance you need to stop** after you see a problem:

> **Thinking distance + Braking distance = Stopping distance**

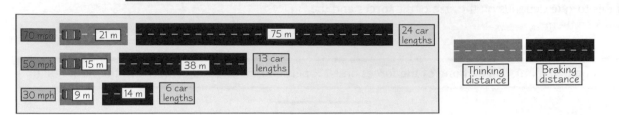

In an exam you might need to list factors that affect the thinking and braking distances.

thinking distance = speed × reaction time

Reaction time is increased by **tiredness, alcohol** or other **drug** use, **illness, distractions** such as noisy children and Wayne's World-style headbanging.

Braking distance depends on the **braking force, friction** between the tyres and the road, the **mass** and the **speed**.

Braking force is reduced by **reduced friction** between the brakes and the wheels (**worn** or **badly adjusted brakes**).

Friction between the tyres and the road is reduced by **wet** or **icy** roads, **leaves or dirt** on the road, **worn-out tyre treads**, etc.

Mass is affected by the size of the car and what you put in it.

Car Safety Features are Usually Designed to Slow You Down Gradually

Modern cars have **safety features** built in. Many of them make use of the idea of slowing the collision down so it **takes you longer to stop**, so your **deceleration is less** and there is **less force** on you.

Safety features you need to know about are:

1) **Seatbelts** keep you in your seat and also 'give' a little so that you're brought to a stop over a longer time.

2) **Airbags** inflate when you have a collision and are big and squishy so they stop you hitting hard things and slow you down gradually. (More about airbags and how they work on the next page.)

3) **Crumple zones** at the front and back of the car are designed to give way more easily and absorb some of the energy of the collision.

4) **Safety cages** are designed to prevent the area around the occupants of the car from being crushed in.

Example

Giles's car bumps into the back of a stationary bus. The car was travelling at 2 ms^{-1} and comes to a stop in 0.2 s. Giles was wearing his seatbelt and takes 0.8 s to stop. The mass of the car is 1000 kg and Giles's mass is 75 kg.

a) Find the decelerations of Giles and the car.

b) Calculate the average force acting on Giles during the accident.

c) Work out the average force that would have acted on Giles if he had stopped in as short a time as the car.

    a)   Use $v = u + at$:
        For the car: $u = 2$ ms^{-1}, $v = 0$, $t = 0.2$ s
        Which gives: $0 = 2 + 0.2a \Rightarrow 0.2a = -2 \Rightarrow a = -10$ ms^{-2} so the **deceleration = 10 ms^{-2}**
        For Giles: $u = 2$ ms^{-1}, $v = 0$, $t = 0.8$ s
        Which gives: $0 = 2 + 0.8a \Rightarrow 0.8a = -2 \Rightarrow a = -2.5$ ms^{-2} so the **deceleration = 2.5 ms^{-2}**

    b)   Use $F = ma = 75 \times 2.5 = $ **187.5 N**

    c)   Use $F = ma$ again, but with 10 ms^{-2} instead of 2.5 ms^{-2}: $F = ma = 75 \times 10 = $ **750 N**

Car Safety

Airbags are Triggered by Rapid Deceleration

It's pretty hard going trying to get an airbag to go off when you need it — and not end up having a face full of airbag every time you stop at traffic lights. Here's how they work...

1) All airbags are **triggered to inflate using** sensors that detect the **rapid deceleration** of a car in a crash.

2) Most cars use a microchip **accelerometer** — where **rapid deceleration** changes the **capacitance** of part of the microchip. This change can be detected by the microchip's electronics, which send an "inflate now" signal to the airbag modules in the car.

3) This kicks off a rapid **chemical reaction** that produces a load of inert gas to inflate the air bag.

4) It's a lot to do in a very **short space of time** if it's going to get to your head before you get up close and personal with a steering wheel. Airbags inflate in less than a tenth of a second once triggered.

5) As soon as they're inflated, the airbags begin to deflate as gas escapes through flaps in the fabric.

GPS Devices Find Where You Are Using Trilateration

Some cars are fitted with glitzy global positioning systems (**GPS**) to help you find out **where** you are. They do this using a process called **trilateration**...

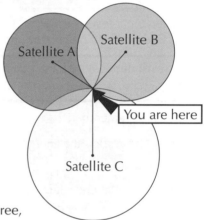

1) The GPS in your car receives signals from at least **three satellites**, each transmitting their **location** and the **time** the signal was sent.

2) As the signals take a short amount of time to reach the GPS, there is a short **delay** between the time sent and the time each signal is received. The further away the satellite, the longer it takes the signal to get to your car.

3) By knowing the **time delay** for each satellite signal, the GPS can calculate the **distance** to each satellite. You then know that you must be somewhere on the surface of a sphere that's centred on that satellite.

4) If you know the distances to **three** satellites, you must be at the point where all three spheres **meet**. Clever, huh?

5) GPS systems actually use at least **four** satellites to locate you. You only need three, but the more satellites the more **accurately** you can know your position.

Practice Questions

Q1   What equation can you use to work out the force you experience during a collision?

Q2   What factor affects both thinking distance and braking distance?

Q3   Describe how airbags are triggered to inflate when a car is in a collision.

Exam Questions

Q1   Sarah sees a cow step into the road 30 m ahead of her. Sarah's reaction time is 0.5 s. She is travelling at 20 ms^{-1}.
Her maximum braking force is 10 000 N and her car (with her in it) has a mass of 850 kg.
(a) How far does she travel before applying her brakes?   [2 marks]
(b) Calculate Sarah's braking distance. Assume she applies the maximum braking force until she stops.   [3 marks]
(c) Does Sarah hit the cow? Justify your answer with a suitable calculation.   [1 mark]

Q2   In a crash test a car slams into a solid barrier at 20 ms^{-1}. The car comes to a halt in 0.1 s. The crash test
dummy goes through the windscreen and hits the barrier at a speed of 18 ms^{-1} and then also comes to a
stop in 0.1 s. The mass of the car is 900 kg and the mass of the dummy is 50 kg.
(a) Calculate the forces on the car and the dummy as they are brought to a stop.   [4 marks]
(b) The car is modified to include crumple zones and an airbag.
    Explain what difference this will make and why.   [3 marks]

Crumple zone — the heap of clothes on my bedroom floor...

Being safe in a car is mainly common sense — don't drive if you're ill, drunk or just tired and don't drive a car with dodgy brakes. But you still need to cope with exam questions, so don't go on till you're sure you know this all off by heart.

Work and Power

As everyone knows, work in Physics isn't like normal work. It's harder. Work also has a specific meaning that's to do with movement and forces. You'll have seen this at GCSE — it just comes up in more detail for AS level.

Work is Done Whenever Energy is Transferred

This table gives you some examples of **work being done** and the **energy changes** that happen.

1) Usually you need a force to move something because you're having to **overcome another force**.

2) The thing being moved has **kinetic energy** while it's **moving**.

3) The kinetic energy is transferred to **another form of energy** when the movement stops.

ACTIVITY	WORK DONE AGAINST	FINAL ENERGY FORM
Lifting up a box.	gravity	gravitational potential energy
Pushing a chair across a level floor.	friction	heat
Pushing two magnetic north poles together.	magnetic force	magnetic energy
Stretching a spring.	stiffness of spring	elastic potential energy

The word **'work'** in Physics means the **amount of energy transferred** from one form to another when a force causes a movement of some sort.

Work = Force × Distance

When a car tows a caravan, it applies a force to the caravan to moves it to where it's wanted.
To **find out** how much **work** has been **done**, you need to use the **equation**:

> **work done (W) = force causing motion (F) × distance moved (s)**
>
> ...where **W** is measured in joules (J), **F** is measured in newtons (N) and **s** is measured in metres (m).

Points to remember:

1) **Work** is the **energy** that's been **changed** from one form to another — it's not necessarily the **total** energy. E.g. moving a book from a low shelf to a higher one will increase its gravitational potential energy, but it had some potential energy to start with. Here, the **work done** would be the **increase** in potential energy, **not the total** potential energy.

2) Remember the distance needs to be measured in metres — if you have **distance in centimetres or kilometres**, you need to **convert** it to metres first.

3) The force **F** will be a **fixed** value in any calculations, either because it's **constant** or because it's the **average** force.

4) The equation assumes that the **direction of the force** is the **same** as the **direction of movement**.

5) The equation gives you the **definition** of the joule (symbol J):
   'One joule is the work done when a force of 1 newton moves an object through a distance of 1 metre'.

The Force isn't always in the Same Direction as the Movement

Sometimes the **direction of movement** is **different** from the **direction of the force**.

Example

1) To **calculate the work done** in a situation like the one on the right, you need to consider the **horizontal** and **vertical components** of the **force**.

2) The only **movement** is in the **horizontal** direction. This means the **vertical force** is not causing any motion (and hence not doing any work) — it's just **balancing** out some of the **weight**, meaning there's a **smaller reaction force**.

direction of force on sledge

rosebud

direction of motion

3) The horizontal force is causing the motion — so to **calculate** the **work done**, this is the **only force** you need to consider. Which means we get:

$$W = Fs \cos \theta$$

Where θ is the **angle** between the **direction of the force** and the **direction of motion**. See page 22 for more on resolving forces.

F

θ

$F \cos \theta$

Direction of motion

Work and Power

Power = Work Done per Second

Power means many things in everyday speech, but in physics (of course!) it has a special meaning. Power is the **rate of doing work** — in other words it is the **amount of energy transformed** from one form to another **per second.**
You **calculate power** from this equation:

> **Power (P) = work done (W) / time (t)**
> ...where **P** is measured in watts (W), **W** is measured in joules (J) and **t** is measured in seconds (s)

The **watt** (symbol W) is defined as a **rate of energy transfer** equal to **1 joule per second** (Js⁻¹).
Yep, that's another **equation and definition** for you to **learn**.

Power is also Force × Velocity (P = Fv)

Sometimes, it's **easier** to use **this version** of the power equation. This is how you get it:

1) You **know** $P = W/t$.
2) You also **know** $W = Fs$, which gives $P = Fs/t$.
3) But $v = s/t$, which you can substitute into the above equation to give $P = Fv$.
4) It's easier to use this if you're given the **speed** in the question.
   Learn this equation as a **shortcut** to link **power** and **speed**.

Example

A car is travelling at a speed of $10\,ms^{-1}$ and is kept going against the frictional force by a driving force of $500\,N$ in the direction of motion. Find the power supplied by the engine to keep the car moving.

Use the shortcut $P = Fv$, which gives:
$P = 500 \times 10 = 5000\,W$

If the force and motion are in different directions, you can replace F with $F \cos \theta$ to get: $\boxed{P = Fv \cos \theta}$

You **aren't** expected to **remember** this equation, but it's made up of bits that you **are supposed to know**, so be ready for the possibility of calculating **power** in a situation where the **direction of the force and direction of motion are different**.

Practice Questions

Q1 Write down the equation used to calculate work if the force and motion are in the same direction.

Q2 Write down the equation for work if the force is at an angle to the direction of motion.

Q3 Write down the equations relating (i) power and work and (ii) power and speed.

Exam Questions

Q1 A traditional narrowboat is drawn by a horse walking along the towpath.
The horse pulls the boat at a constant speed between two locks which are
1500 m apart. The tension in the rope is 100 N at 40° to the direction of motion.

(a) How much work is done on the boat? [2 marks]
(b) The boat moves at 0.8 ms⁻¹. Calculate the power supplied to the boat in the direction of motion. [2 marks]

Q2 A motor is used to lift a 20 kg load a height of 3 m. (Take g = 9.81 Nkg⁻¹.)

(a) Calculate the work done in lifting the load. [2 marks]
(b) The speed of the load during the lift is 0.25 ms⁻¹. Calculate the power delivered by the motor. [2 marks]

Work — there's just no getting away from it...

Loads of equations to learn. Well, that's what you came here for, after all. Can't beat a good bit of equation-learning, as I've heard you say quietly to yourself when you think no one's listening. Aha, can't fool me. Ahahahahahahahahahahahaha.

Conservation of Energy

Energy can never be *lost*. I repeat — *energy* can *never* be lost. Which is basically what I'm about to take up two whole pages saying. But that's, of course, because you need to do exam questions on this as well as understand the principle.

Learn the **Principle** of **Conservation** of **Energy**

The **principle of conservation of energy** says that:

> Energy **cannot be created** or **destroyed**. Energy **can be transferred** from one form to another but the total amount of energy in a closed system will not change.

Example

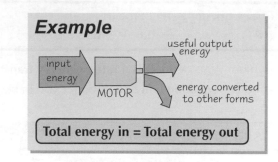

input energy ➡ MOTOR ➡ useful output energy / energy converted to other forms

Total energy in = Total energy out

You need it for **Questions** about **Kinetic** and **Potential Energy**

The principle of conservation of energy nearly always comes up when you're doing questions about changes between kinetic and potential energy.

A quick reminder:

1) **Kinetic energy** is energy of anything **moving**, which you work out from $E_k = \frac{1}{2}mv^2$, where v is the velocity it's travelling at and m is its mass.

2) There are **different types of potential energy** — e.g. gravitational and elastic.

3) **Gravitational potential energy** is the energy something gains if you lift it up. You work it out using: $\Delta E_p = mg\Delta h$, where m is the mass of the object, Δh is the height it is lifted and g is the gravitational field strength (9.81 Nkg⁻¹ on Earth).

4) **Elastic potential energy** (elastic stored energy) is the energy you get in, say, a stretched rubber band or spring. You work this out using $E = \frac{1}{2}ke^2$, where e is the extension of the spring and k is the stiffness constant.

Examples

These pictures show you three **examples** of changes between kinetic and potential energy.

1) As Becky throws the **ball upwards, kinetic energy** is converted into **gravitational potential energy**. When it **comes down** again, that **gravitational potential** energy is **converted back** into **kinetic** energy.

2) As Dominic goes **down the slide, gravitational potential energy** is converted to **kinetic energy**.

3) As Simon bounces upwards from the trampoline, **elastic potential energy** is converted to **kinetic energy**, to **gravitational potential energy**. As he comes back down again, that **gravitational potential** energy is **converted back** to **kinetic** energy, to **elastic potential** energy, and so on.

> In **real life** there are also **frictional forces** — Simon would have to use some **force** from his **muscles** to keep **jumping** to the **same height** above the trampoline each time. Each time the trampoline **stretches**, some **heat** is generated in the trampoline material. You're usually told to **ignore friction** in exam questions — this means you can **assume** that the **only forces** are those that provide the **potential or kinetic energy** (in this example that's **Simon's weight** and the **tension** in the springs and trampoline material).
> If you're ignoring friction, you can say that the **sum of the kinetic and potential energies is constant**.

Conservation of Energy

Use Conservation of Energy to Solve Problems

You need to be able to **use** conservation of mechanical energy (change in potential energy = change in kinetic energy) to solve problems. The classic example is the **simple pendulum**.

In a simple pendulum, you assume that all the mass is in the **bob** at the end.

Example

A simple pendulum has a mass of 700 g and a length of 50 cm. It is pulled out to an angle of 30° from the vertical.

(a) Find the gravitational potential energy stored in the pendulum bob.

Start by drawing a diagram.

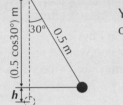

You can work out the increase in height, **h**, of the end of the pendulum using trig.

Gravitational potential energy = **mgh**
$$= 0.7 \times 9.81 \times (0.5 - 0.5\cos30°)$$
$$= 0.46 \text{ J}$$

(b) The pendulum is released. Find the maximum speed of the pendulum bob as it passes the vertical position.

To find the *maximum* speed, assume no air resistance, then **mgh** = ½**mv²**.

So $\dfrac{1}{2}mv^2 = 0.46$

rearrange to find $v = \sqrt{\dfrac{2 \times 0.46}{0.7}} = 1.15 \text{ ms}^{-1}$

OR

Cancel the **m**s and rearrange to give:
$$v^2 = 2gh$$
$$= 2 \times 9.81 \times (0.5 - 0.5\cos30°)$$
$$= 1.31429...$$
$$v = 1.15 \text{ ms}^{-1}$$

You could be asked to apply this stuff to just about any situation in the exam. **Rollercoasters** are a bit of a favourite.

Practice Questions

Q1 State the principle of conservation of energy.

Q2 What are the equations for calculating kinetic energy and gravitational potential energy?

Q3 Show that, if there's no air resistance and the mass of the string is negligible, the speed of a pendulum is independent of the mass of the bob.

Exam Questions

Q1 A skateboarder is on a half-pipe. He lets the board run down one side of the ramp and up the other. The height of the ramp is 2 m. Take **g** as 9.81 Nkg⁻¹.

(a) If you assume that there is no friction, what would be his speed at the lowest point of the ramp? [3 marks]

(b) How high will he rise up the other side? [1 mark]

(c) Real ramps are not frictionless, so what must the skater do to reach the top on the other side? [1 mark]

Q2 A 20 g rubber ball is released from a height of 8 m. (Assume that the effect of air resistance is negligible.)

(a) Find the kinetic energy of the ball just before it hits the ground. [2 marks]

(b) The ball strikes the ground and rebounds to a height of 6.5 m. How much energy is converted to heat and sound in the impact with the ground? [2 marks]

Energy is never lost — it just sometimes prefers the scenic route...

Remember to check your answers — I can't count the number of times I've forgotten to square the velocities or to multiply by the ½... I reckon it's definitely worth the extra minute to check.

Efficiency and Sankey Diagrams

Energy, it seems, is pretty reliable stuff — whatever you do, you can't create or destroy it — it'll always be there.
But, like homework and small children, it's very easy to lose — you know it's there somewhere, you just can't find it.

All Energy Transfers Involve Losses

You saw on the last page that **energy can never be created or destroyed**. But whenever **energy** is **converted** from one form to another, some is always **'lost'**. It's still there (i.e. it's **not destroyed**) — it's just not in a form you can **use**.

Most often, **energy** is lost as **heat** — e.g. **computers** and **TVs** are always **warm** when they've been on for a while. In fact, **no device** (except possibly a heater) is ever **100% efficient** (see below) because some energy is **always** lost as **heat**. (You want heaters to give out heat, but in other devices the heat loss isn't useful.) Energy can be **lost** in other forms too (e.g. **sound**) — the important thing is the lost energy **isn't** in a **useful** form and you **can't** get it back.

Efficiency is the Ratio of Useful Energy Output to Total Energy Input

Efficiency is one of those words we use all the time, but it has a **specific meaning** in Physics. It's a measure of how well a **device** converts the **energy** you put **in** into the energy you **want** it to give **out**. So, a device that **wastes** loads of **energy** as heat and sound has a really **low efficiency**, and vice versa.

There's a nice little **equation** to find the **efficiency** of a device:

$$\text{Efficiency} = \frac{\text{useful output energy}}{\text{total input energy}} \times 100\%$$

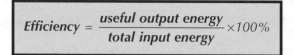

Energy, as always, is measured in joules (J). Efficiency has no units because it's a ratio.

Some questions will be kind and **give you** the **useful output energy** — others will tell you how much is **wasted**. You just have to **subtract** the **wasted energy** from the **total input energy** to find the **useful output energy**, so it's not too tricky if you keep your wits about you.

Sankey Diagrams Show Energy Input and Output

Sankey diagrams (or energy transformation diagrams) are a way of showing how much of the **input energy** is being **usefully employed** compared with how much is being **wasted**.

For example, the **Sankey diagram** for an **electric motor** is shown below.

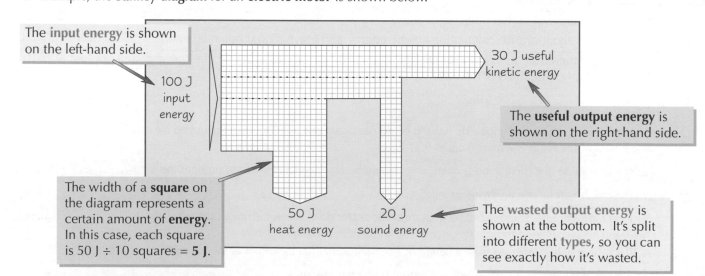

The **input energy** is shown on the left-hand side.

100 J input energy

30 J useful kinetic energy

The **useful output energy** is shown on the right-hand side.

The width of a **square** on the diagram represents a certain amount of **energy**. In this case, each square is 50 J ÷ 10 squares = **5 J**.

50 J heat energy

20 J sound energy

The **wasted output energy** is shown at the bottom. It's split into different **types**, so you can see exactly how it's wasted.

The **useful** thing about **Sankey diagrams** is that you can see what's going on at a glance — there's always **one** big **arrow** going in (**input energy**) and a **few** smaller ones **coming out** (**output energy**). The **width** of the arrows tells you how much **energy** is in each form — the **thicker** the arrow, the **greater** the amount of **energy**.

You can even use **Sankey diagrams** to work out the **efficiency** of the device — read off the **total input energy** and **useful output energy**, then substitute them in the **equation** above.

Efficiency and Sankey Diagrams

Drawing Sankey Diagrams is Easy — if You Take it Step by Step

In the exam, you could be asked to **interpret** a Sankey diagram — or you could be asked to **draw** one for yourself. Take it **step by step** and you'll be fine.

1) Find the **total input energy**, the **useful output energy** and the **amount of energy wasted** in each different form. You might be given **all** these values — or you might be given **some** and have to **add** or **subtract** to find the rest.

2) Choose your **scale**. It's best (and easiest to draw) if you can represent **all** the energy values by a **whole number** of squares. It might sound obvious, but you also want your **diagram** to be a **sensible size**.

3) Work out **how many squares** will represent each energy value: **energy ÷ energy per square = number of squares**.

4) **Draw** the **arrow** for the **total input energy** — make sure it's the right number of squares or it'll all go wrong.

5) **Split** the **input energy** into all the different **outputs** — the **useful energy** output should go at the **top**.

6) **Draw** the **output arrows** — **useful energy** should go **straight across**, and **wasted energy** should point **downwards**.

7) **Label** all the **arrows** so it's clear what each bit represents.

Example

A car uses 280 MJ of chemical energy to travel 100 km. During this time, 180 MJ are wasted as heat energy and 30 MJ are wasted as sound energy. Draw a Sankey diagram for the car, then calculate its efficiency.

Start by listing the energy values: **Total input energy = 280 MJ**, **Energy wasted as heat = 180 MJ**, **Energy wasted as sound = 30 MJ**, **Useful output energy** = 280 – 180 – 30 = **70 MJ**.

All the values divide by 10, so use a **scale** of **1 square : 10 MJ**.

Draw the **input arrow**.
280 MJ ÷ 10 = **28 squares**

Split it into the different **outputs**.

Draw the remaining **arrows**, then **label** them all.

Work out the efficiency:

$$\text{Efficiency} = \frac{\text{useful output energy}}{\text{total input energy}} \times 100\%$$

$$= \frac{70}{280} \times 100 = \textbf{25\%}$$

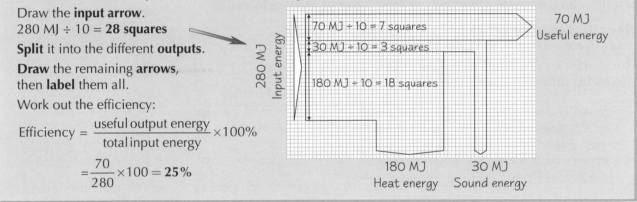

70 MJ ÷ 10 = 7 squares
30 MJ ÷ 10 = 3 squares
180 MJ ÷ 10 = 18 squares

280 MJ Input energy

70 MJ Useful energy

180 MJ Heat energy   30 MJ Sound energy

Practice Questions

Q1　Why can a device never be 100% efficient?
Q2　What is the equation for efficiency?
Q3　Calculate the efficiency of a device that wastes 65 J for every 140 J of input energy.
Q4　What are Sankey diagrams used for?

Exam Question

Q1　The figure on the right is a Sankey diagram for a certain design of wind turbine. A second design of wind turbine produces 30 kJ of electrical energy for every 125 kJ of input kinetic energy. It wastes 70 kJ as sound, and the rest as heat.

60 kJ Input energy

15 kJ Electrical energy

30 kJ Sound energy   15 kJ Heat energy

　　(a) Draw and label a Sankey diagram for the second design of wind turbine.　[4 marks]

　　(b) Which design of wind turbine is more efficient? By how much?　[3 marks]

Sankey diagrams are a physicist's best friend...

I'm quite a fan of Sankey diagrams — you take a load of numbers that don't mean very much, draw a nice Sankey diagram, and hey presto, you can see what's going on. Whether you agree or not, make sure you can draw and interpret them.

Hooke's Law

Hooke's law doesn't apply to all materials, and only works for the rest up to a point, but it's still pretty handy.

Hooke's Law Says that Extension is Proportional to Force

If a **metal wire** is supported at the top and then a weight attached to the bottom, it **stretches**.
The weight pulls down with force **F**, producing an equal and opposite force at the support.

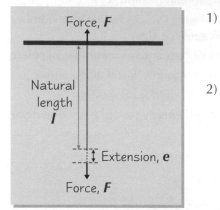

1) **Robert Hooke** discovered in 1676 that the extension of a stretched wire, **e**, is proportional to the load or force, **F**.

   This relationship is now called **Hooke's law**.

2) Hooke's law can be written:

$$F = ke$$

Where **k** is a constant that depends on the material being stretched. **k** is called the **stiffness constant**.

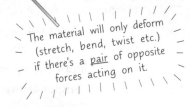

The material will only deform (stretch, bend, twist etc.) if there's a <u>pair</u> of opposite forces acting on it.

I'm a bit irrelevant on this page — bungee ropes don't obey Hooke's Law... Do you think I need to get out more?

Hooke's law Also Applies to Springs

A metal spring also changes length when you apply a **pair of opposite forces**.

1) The **extension** or **compression** of a spring is **proportional** to the **force** applied — so Hooke's law applies.

2) For springs, **k** in the formula **F = ke** is usually called the **spring stiffness** or **spring constant**.

> Hooke's law works just as well for **compressive** forces as **tensile** forces. For a spring, **k** has the **same value** whether the forces are tensile or compressive (that's not true for all materials).

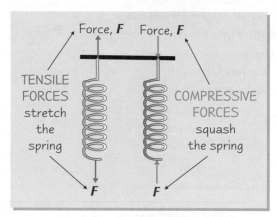

TENSILE FORCES stretch the spring

COMPRESSIVE FORCES squash the spring

Hooke's law Stops Working when the Load is Great Enough

There's a **limit** to the force you can apply for Hooke's law to stay true.

1) The graph shows load against extension for a **typical metal wire**.

2) The first part of the graph shows Hooke's law being obeyed — there's a **straight-line relationship** between **load** and **extension**.

3) When the load becomes great enough, the graph starts to **curve**. The point marked E on the graph is called the **elastic limit**.

4) If you increase the load past the elastic limit, the material will be **permanently stretched**. When all the force is removed, the material will be **longer** than at the start.

5) **Metals** generally obey Hooke's law up to the limit of proportionality (see p.40), which is very near the elastic limit.

6) Be careful — there are some materials, like **rubber**, that only obey Hooke's law for **really small** extensions.

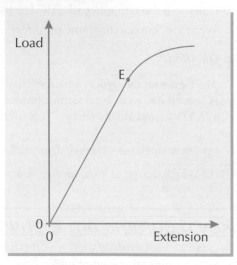

Hooke's Law

So basically...

A Stretch can be *Elastic* or *Plastic*

Elastic

If a **deformation** is **elastic**, the material returns to its **original shape** once the forces are removed.

1) When the material is put under **tension**, the **atoms** of the material are **pulled apart** from one another.
2) Atoms can **move** small distances relative to their **equilibrium positions**, without actually changing position in the material.
3) Once the **load** is **removed**, the atoms **return** to their **equilibrium** distance apart.

For a metal, elastic deformation happens as long as **Hooke's law** is obeyed.

Plastic

If a deformation is **plastic**, the material is **permanently stretched**.

1) Some atoms in the material move position relative to one another.
2) When the load is removed, the **atoms don't return** to their original positions.

A metal stretched **past its elastic limit** shows plastic deformation.

Practice Questions

Q1 State Hooke's law.

Q2 Define tensile forces and compressive forces.

Q3 Explain what is meant by the elastic limit of a material.

Q4 From studying the force-extension graph for a material as it is loaded and unloaded, how can you tell:
(a) if Hooke's law is being obeyed,
(b) if the elastic limit has been reached?

Q5 What is plastic behaviour of a material under load?

Exam Questions

Q1 A metal guitar string stretches 4.0 mm when a 10 N force is applied.

(a) If the string obeys Hooke's law, how far will the string stretch with a 15 N force? [1 mark]

(b) Calculate the stiffness constant for this string in Nm^{-1}. [2 marks]

(c) The string is tightened beyond its elastic limit. What would be noticed about the string? [1 mark]

Q2 A rubber band is 6.0 cm long. When it is loaded with 2.5 N, its length becomes 10.4 cm. Further loading increases the length to 16.2 cm when the force is 5.0 N.

Does the rubber band obey Hooke's law when the force on it is 5.0 N? Justify your answer with a suitable calculation. [2 marks]

Sod's Law — if you don't learn it, it'll be in the exam...

Three things you didn't know about Robert Hooke — he was the first person to use the word 'cell' (in terms of biology, not prisons), he helped Christopher Wren with his designs for St. Paul's Cathedral and no-one knows quite what he looked like. I'd like to think that if I did all that stuff, then someone would at least remember what I looked like — poor old Hooke.

Stress and Strain

How much a material stretches for a particular applied force depends on its dimensions.
If you want to compare the properties of two different materials, you need to use stress and strain instead.
A stress-strain graph is the same for any sample of a particular material — the size of the sample doesn't matter.

A Stress Causes a Strain

A material subjected to a pair of **opposite forces** might **deform**, i.e. **change shape**. If the forces **stretch** the material, they're **tensile**. If the forces **squash** the material, they're **compressive**.

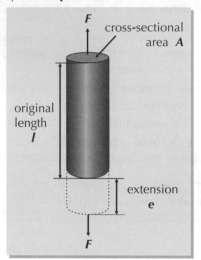

1) **Tensile stress** is defined as the **force applied**, **F**, divided by the **cross-sectional area**, **A**:

$$\text{stress} = \frac{F}{A}$$

The **units** of stress are **Nm⁻²** or pascals, **Pa**.

2) **Tensile strain** is defined as the **change in length**, i.e. the **extension**, divided by the **original length** of the material:

$$\text{strain} = \frac{e}{l}$$

Strain has **no units** — it's just a **number**.

3) It doesn't matter whether the forces producing the **stress** and **strain** are **tensile** or **compressive** — the **same equations** apply. The only difference is that you tend to think of **tensile** forces as **positive**, and **compressive** forces as **negative**.

A Stress Big Enough to Break the Material is Called the Breaking Stress

As a greater and greater tensile **force** is applied to a material, the **stress** on it **increases**.

1) The effect of the **stress** is to start to **pull** the **atoms apart** from one another.

2) Eventually the stress becomes **so great** that atoms **separate completely**, and the **material breaks**. This is shown by point **B** on the graph. The stress at which this occurs is called the **breaking stress**.

3) The point marked **UTS** on the graph is called the **ultimate tensile stress**. This is the **maximum stress** that the material can withstand.

4) **Engineers** have to consider the **UTS** and **breaking stress** of materials when designing a **structure**.

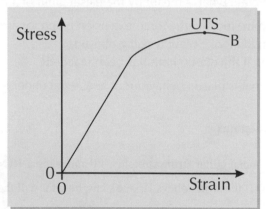

Elastic Strain Energy is the Energy Stored in a Stretched Material

When a material is **stretched**, **work** has to be done in stretching the material.

1) **Before** the **elastic limit**, **all** the **work done** in stretching is **stored** as **potential energy** in the material.

2) This stored energy is called **elastic strain energy**.

3) On a **graph** of **force against extension**, the elastic strain energy is given by the **area under the graph**.

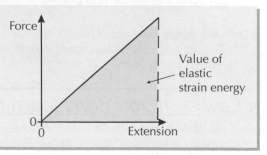

Stress and Strain

You can Calculate the Energy Stored in a Stretched Wire

Provided a material obeys Hooke's law, the **potential energy** stored inside it can be **calculated** quite easily.

1) The work done on the wire in stretching it is equal to the energy stored.

2) **Work done** equals **force × displacement**.

3) However, the **force** on the material **isn't constant**. It rises from zero up to force **F**.
To calculate the **work done**, use the average force between zero and **F**, i.e. ½**F**.

> work done = ½**F** × **e**

4) Then the **elastic strain energy**, **E**, is:

> **E** = ½**Fe**

This is the triangular area under the force-extension graph — see previous page.

5) Because Hooke's law is being obeyed, **F** = **ke**,
which means **F** can be replaced in the equation to give:

> **E** = ½**ke²**

6) If the material is stretched beyond the **elastic limit**, some work is done separating atoms.
This will **not** be **stored** as strain energy and so isn't available when the force is released.

Practice Questions

Q1 Write a definition for tensile stress.

Q2 Explain what is meant by the tensile strain on a material.

Q3 What is meant by the breaking stress of a material?

Q4 How can the elastic strain energy be found from the force against extension graph of a material under load?

Q5 The work done is usually calculated as force multiplied by displacement.
Explain why the work done in stretching a wire is ½*Fe*.

Exam Questions

Q1 A steel wire is 2.00 m long. When a 300 N force is applied to the wire, it stretches 4.0 mm.
The wire has a circular cross-section with a diameter of 1.0 mm.

    (a) What is the cross-sectional area of the wire? [1 mark]

    (b) Calculate the tensile stress in the wire. [1 mark]

    (c) Calculate the tensile strain of the wire. [1 mark]

Q2 A copper wire (which obeys Hooke's law) is stretched by 3.0 mm when a force of 50 N is applied.

    (a) Calculate the stiffness constant for this wire in Nm^{-1}. [2 marks]

    (b) What is the value of the elastic strain energy in the stretched wire? [1 mark]

Q3 A pinball machine contains a spring which is used to fire a small, 12 g metal ball to start the game.
The spring has a stiffness constant of 40.8 Nm^{-1}. It is compressed by 5 cm and then released to fire the ball.

Calculate the maximum possible speed of the ball. [4 marks]

UTS a laugh a minute, this stuff...

Here endeth the proper physics for this section — the rest of it's materials science (and I don't care what your exam boards say). It's all a bit "useful" for my liking. Calls itself a physics course... grumble... grumble... wasn't like this in my day... But to be fair — some of it's quite interesting, and there are some pretty pictures coming up on page 66.

The Young Modulus

Busy chap, Thomas Young. He did this work on tensile stress as something of a sideline. Light was his main thing.
He proved that light behaved like a wave, explained how we see in colour and worked out what causes astigmatism.

The **Young Modulus** is Stress ÷ Strain

When you apply a **load** to stretch a material, it experiences a **tensile stress** and a **tensile strain**.

1) Up to a point called the **limit of proportionality** (see p.40), the stress and strain of a material are proportional to each other.

2) So below this limit, for a particular material, stress divided by strain is a constant. This constant is called the **Young modulus**, *E*.

$$E = \frac{\text{tensile stress}}{\text{tensile strain}} = \frac{F/A}{e/l} = \frac{Fl}{eA}$$

Where, *F* = force in N, *A* = cross-sectional area in m², *l* = initial length in m and *e* = extension in m.

3) The **units** of the Young modulus are the same as stress (**Nm⁻²** or pascals), since strain has no units.

4) The Young modulus is used by **engineers** to make sure their materials can withstand sufficient forces.

To **Find** the Young Modulus, You need a **Very Long Wire**

This is the experiment you're most likely to do in class:

Mum moment: if you're doing this experiment, wear safety goggles — if the wire snaps, it could get very messy...

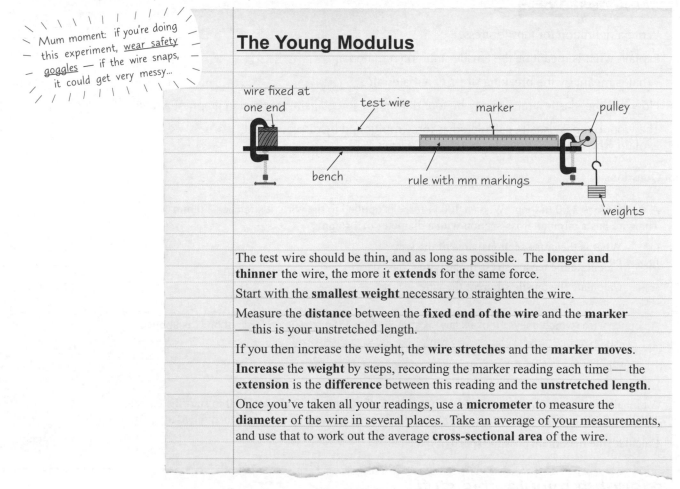

The Young Modulus

wire fixed at one end · test wire · marker · pulley · bench · rule with mm markings · weights

The test wire should be thin, and as long as possible. The **longer and thinner** the wire, the more it **extends** for the same force.

Start with the **smallest weight** necessary to straighten the wire.

Measure the **distance** between the **fixed end of the wire** and the **marker** — this is your unstretched length.

If you then increase the weight, the **wire stretches** and the **marker moves**.

Increase the **weight** by steps, recording the marker reading each time — the **extension** is the **difference** between this reading and the **unstretched length**.

Once you've taken all your readings, use a **micrometer** to measure the **diameter** of the wire in several places. Take an average of your measurements, and use that to work out the average **cross-sectional area** of the wire.

The other standard way of measuring the Young modulus in the lab is using **Searle's apparatus**. This is a bit more accurate, but it's harder to do and the equipment's more complicated.

The Young Modulus

Use a **Stress-Strain Graph** to Find **E**

You can plot a **graph** of **stress against strain** from your results.

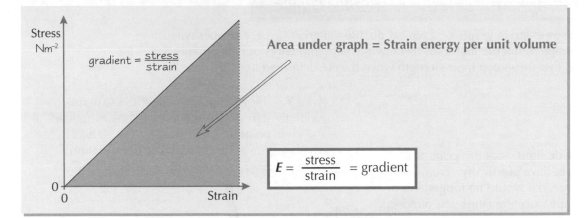

1) The **gradient** of the graph gives the Young modulus, **E**.

2) The **area under the graph** gives the **strain energy** (or energy stored) per unit volume i.e. the energy stored per 1 m³ of wire.

3) The stress-strain graph is a **straight line** provided that Hooke's law is obeyed, so you can also calculate the energy per unit volume as:

$$\text{energy} = \tfrac{1}{2} \times \textbf{stress} \times \textbf{strain}$$

Practice Questions

Q1 Define the Young modulus for a material.

Q2 What are the units for the Young modulus?

Q3 Explain why a thin test wire is used to find the Young modulus.

Q4 What is given by the area contained under a stress-strain graph?

Exam Questions

Q1 A steel wire is stretched elastically. For a load of 80 N, the wire extends by 3.6 mm. The original length of the wire was 2.50 m and its average diameter is 0.6 mm.

(a) Calculate the cross-sectional area of the wire in m². [1 mark]

(b) Find the tensile stress applied to the wire. [1 mark]

(c) Calculate the tensile strain of the wire. [1 mark]

(d) What is the value of the Young modulus for steel? [1 mark]

Q2 The Young modulus for copper is 1.3×10^{11} Nm⁻².

(a) If the stress on a copper wire is 2.6×10^{8} Nm⁻², what is the strain? [2 marks]

(b) If the load applied to the copper wire is 100 N, what is the cross-sectional area of the wire? [1 mark]

(c) Calculate the strain energy per unit volume for this loaded wire. [1 mark]

Learn that experiment — it's important...

Getting back to the good Dr Young... As if ground-breaking work in light, the physics of vision and materials science wasn't enough, he was also a well-respected physician, a linguist and an Egyptologist. He was one of the first to try to decipher the Rosetta stone (he didn't get it right, but nobody's perfect). Makes you feel kind of inferior, doesn't it. Best get learning.

Interpreting Stress-Strain Graphs

Remember that lovely stress-strain graph from page 36? Well, turns out that because materials have different properties, their stress-strain graphs look different too — you need to know the graphs for ductile, brittle and polymeric materials.

Stress-Strain Graphs for **Ductile** Materials **Curve**

The diagram shows a **stress-strain graph** for a typical **ductile** material — e.g. a copper wire.
You can change the **shape** of **ductile materials** by drawing them into **wires** or other shapes.
The important thing is that they **keep their strength** when they're deformed like this.

Point **Y** is the **yield point** — here the material suddenly starts to **stretch** without any extra load. The **yield point** (or yield stress) is the **stress** at which a large amount of **plastic deformation** takes place with a **constant** or **reduced load**.

Point **E** is the **elastic limit** — at this point the material starts to behave **plastically**. From point E onwards, the material would **no longer** return to its **original shape** once the stress was removed.

Point **P** is the **limit of proportionality** — after this, the graph is no longer a straight line but starts to **bend**. At this point, the material **stops** obeying **Hooke's law**, but would still **return** to its **original shape** if the stress was removed.

Before point **P**, the graph is a **straight line** through the **origin**. This shows that the material is obeying **Hooke's law** (page 34).

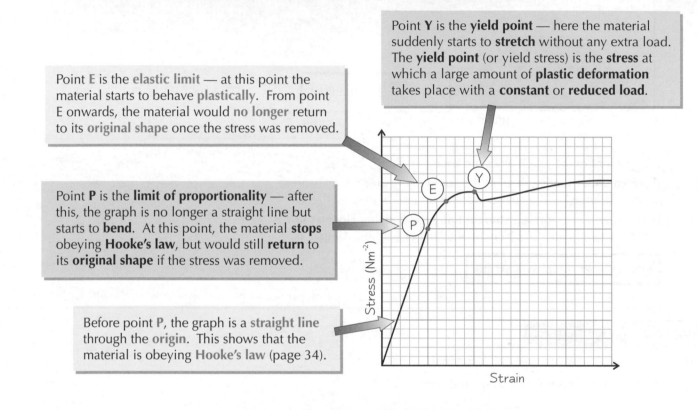

Stress-Strain Graphs for **Brittle** Materials **Don't Curve**

The graph shown below is typical of a **brittle** material.

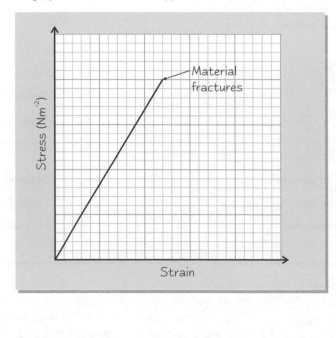

Material fractures

The graph starts the same as the one above — with a **straight line through the origin**. So brittle materials also obey **Hooke's law**.

However, when the **stress** reaches a certain point, the material **snaps** — it doesn't deform plastically.

When **stress** is applied to a brittle material any **tiny cracks** at the material's surface get **bigger** and **bigger** until the material **breaks** completely. This is called **brittle fracture**.

Hooke's law — it's the pirates' code... yarr

Interpreting Stress-Strain Graphs

Rubber and Polythene Are Polymeric Materials

1) The **molecules** that make up **polymeric** (or polymer) **materials** are arranged in **long chains**.

2) They have a **range** of properties, so different polymers have different **stress-strain graphs**.
The diagram below shows two examples — **rubber** and **polythene**.

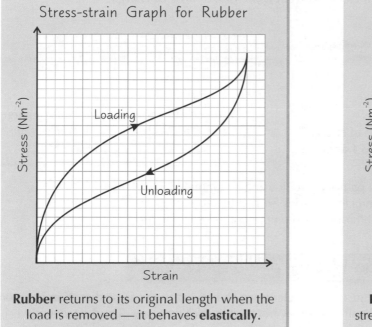

Stress-strain Graph for Rubber

Rubber returns to its original length when the load is removed — it behaves **elastically**.

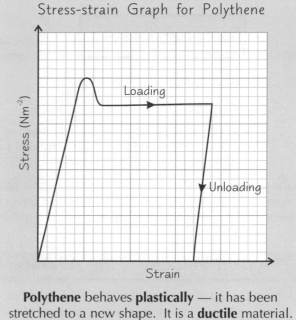

Stress-strain Graph for Polythene

Polythene behaves **plastically** — it has been stretched to a new shape. It is a **ductile** material.

Practice Questions

Q1 Write short definitions of the following terms: ductile, brittle.

Q2 What is the difference between the limit of proportionality and the elastic limit?

Q3 What are polymeric materials?

Q4 Sketch stress-strain graphs of typical ductile, brittle and polymeric materials and describe their shapes.

Exam Questions

Q1 Hardened steel is a hard, brittle form of steel made by heating it up slowly and then quenching it in cold water.

(a) What is meant by the term *brittle*? [2 marks]

(b) Sketch a stress-strain graph for hardened steel. [2 marks]

Q2 An electric cable consists of a copper wire surrounded by polythene insulation.

(a) Sketch stress-strain graphs for the two materials. [3 marks]

(b) Describe two similarities between the behaviour of the two materials under stress. [2 marks]

My sister must be brittle — she's always snapping...

In case you were wondering, I haven't just drawn the graphs on these two pages for fun (though I did enjoy myself) — they're there for you to learn. I find the best way to remember each one is to understand why it has the shape it does — if that sounds too much like hard work, then at least make sure you can describe the shape of all four of them.

Charge, Current and Potential Difference

You wouldn't reckon there was that much to know about electricity... just plug something in, and bosh — electricity.
Ah well, never mind the age of innocence — here are all the gory details...

Current is the Rate of Flow of Charge

The **current** in a **wire** is like **water** flowing in a **pipe**. The **amount** of water that flows depends on the
flow rate and the **time**. It's the same with electricity — **current is the rate of flow of charge**.

$$\Delta Q = I\Delta t \quad \text{or} \quad I = \frac{\Delta Q}{\Delta t}$$

Where ΔQ is the charge in coulombs,
I is the current and Δt is the time taken.

*Remember that conventional current flows from
+ to -, the opposite way from electron flow.*

The Coulomb is the Unit of Charge

One **coulomb** (**C**) is defined as the **amount of charge**
that passes in **1 second** when the **current** is **1 ampere**.

You need to know the elementary charge
too (i.e. the charge on a single electron):
$$e = 1.6 \times 10^{-19} \text{ C}$$

You can measure the current flowing through a part of a circuit using an **ammeter**.
Remember — you always need to attach an ammeter in **series** (so that the current
through the ammeter is the same as the current through the component — see page 54).

The Drift Velocity is the Average Velocity of the Electrons

When **current** flows through a wire, you might imagine the **electrons** all moving in the **same direction** in an orderly
manner. Nope. In fact, they move **randomly** in **all directions**, but tend to **drift** one way. The **drift velocity** is just the
average velocity and it's **much, much less** than the electrons' **actual speed**. (Their actual speed is about 10^6 ms^{-1}!)

The Current Depends on the Drift Velocity

The **current** is given by the equation: $I = nAvq$ *You don't need to derive this for the exam but
you do need to understand what it means.*

where: I = electric current in A v = drift velocity in ms^{-1}
            n = number of charge carriers per m^3 q = charge in C carried by each charge carrier
            A = cross-sectional area in m^2

See what the Equation Means by Changing One Variable at a Time

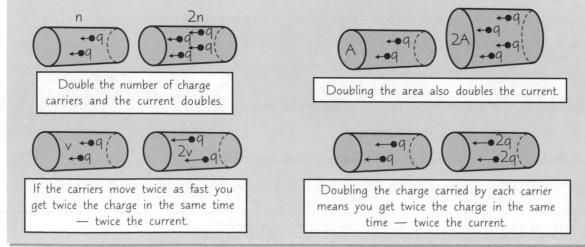

Double the number of charge
carriers and the current doubles.

Doubling the area also doubles the current.

If the carriers move twice as fast you
get twice the charge in the same time
— twice the current.

Doubling the charge carried by each carrier
means you get twice the charge in the same
time — twice the current.

Charge, Current and Potential Difference

Different Materials have Different Numbers of Charge Carriers

1) In a **metal**, the **charge carriers** are **free electrons** — they're the ones from the **outer shell** of each atom. Thinking about the formula $I = nAvq$, there are **loads** of charge carriers, making n **big**. The **drift velocity** only needs to be **small**, even for a **high current**.

2) **Semiconductors** have **fewer charge carriers** than metals, so the **drift velocity** will need to be **higher** if you're going to have the **same current**.

3) A **perfect insulator** wouldn't have **any charge carriers**, so $n = 0$ in the formula and you'd get **no current**. **Real** insulators have a **very small n**.

Charge Carriers in Liquids and Gases are Ions

1) **Ionic crystals** like sodium chloride are **insulators**. Once **molten**, though, the liquid **conducts**. Positive and negative **ions** are the **charge carriers**. The **same thing** happens in an **ionic solution** like copper sulfate solution.

2) **Gases** are **insulators**, but if you apply a **high enough voltage** electrons get **ripped out** of **atoms**, giving you **ions** along a path. You get a **spark**.

Potential Difference is the Energy per Unit Charge

To make electric charge flow through a conductor, you need to do work on it. **Potential difference** (p.d.), or **voltage**, is defined as the **energy converted per unit charge moved**.

$$V = \frac{W}{Q}$$

W is the energy in joules. It's the work you do moving the charge.

Back to the 'water analogy' again. The p.d. is like the pressure that's forcing water along the pipe.

Resistor

6V

Here you do 6 J of work moving each coulomb of charge through the resistor, so the p.d. across it is 6 V. The energy gets converted to heat.

Definition of the Volt

The **potential difference** across a component is **1 volt** when you convert **1 joule** of energy moving **1 coulomb** of charge through the component.

$$1\,V = 1\,J\,C^{-1}$$

Practice Questions

Q1 Describe in words how current and charge are related.
Q2 Define the coulomb.
Q3 Explain what drift velocity is.
Q4 Define potential difference.

Exam Questions

Q1 A battery delivers 4500 C of electric charge to a circuit in 10 minutes. Calculate the average current. [2 marks]

Q2 Copper has 1.0×10^{29} free electrons per m^3. Calculate the drift velocity of the electrons in a copper wire of cross-sectional area $5.0 \times 10^{-6}\,m^2$ when it is carrying a current of 13 A. (electron charge $= 1.6 \times 10^{-19}\,C$) [3 marks]

Q3 An electric motor runs off a 12 V d.c. supply and has an overall efficiency of 75%. Calculate how much electric charge will pass through the motor when it does 90 J of work. [3 marks]

I can't even be bothered to make the current joke...

Talking of currant jokes, I saw this bottle of wine the other day called 'raisin d'être' — 'raison d'être' of course meaning 'reason for living', but spelled slightly different to make 'raisin', meaning 'grape'. Ho ho. Chuckled all the way out of Tesco.

Resistance and Resistivity

"You will be assimilated. Resistance is futile."

Sorry, I couldn't resist it (no pun intended), and I couldn't think of anything useful to write anyway. This resistivity stuff gets a bit more interesting when you start thinking about temperature and light dependence, but for now, just learn this.

Everything *has* Resistance

1) If you put a **potential difference** (p.d.) across an **electrical component**, a **current** will flow.

2) **How much** current you get for a particular **p.d.** depends on the **resistance** of the component.

3) You can think of a component's **resistance** as a **measure** of how **difficult** it is to get a **current** to **flow** through it.

> Mathematically, **resistance** is: $\boxed{R = \dfrac{V}{I}}$
>
> This equation really **defines** what is meant by resistance.

4) **Resistance** is measured in **ohms** (Ω).

> A component has a resistance of **1 Ω** if a **potential difference** of **1 V** makes a **current** of **1 A** flow through it.

Three Things *Determine* Resistance

If you think about a nice, **simple electrical component**, like a **length of wire**, its **resistance** depends on:

1) **Length (***l***).** The **longer** the wire the **more difficult** it is to make a **current flow**.

2) **Area (***A***).** The **wider** the wire the **easier** it will be for the electrons to pass along it.

3) **Resistivity (***ρ***).** This **depends** on the **material**. The **structure** of the material of the wire may make it easy or difficult for charge to flow. In general, resistivity depends on **environmental factors** as well, like **temperature** and **light intensity**.

> The **resistivity** of a material is defined as the **resistance** of a **1m length** with a **1m² cross-sectional area**. It is measured in **ohm metres** (Ωm).

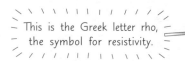 This is the Greek letter rho, the symbol for resistivity.

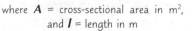 $$\rho = \frac{RA}{l}$$ where **A** = cross-sectional area in m², and **l** = length in m

You will more **usually** see the equation in the **form**: $$R = \rho \frac{l}{A}$$

Typical values for the **resistivity** of **conductors** are **really small**.

For example, the resistivity of **copper** (at 25 °C) is just 1.72×10^{-8} Ωm.

If you **calculate** a **resistance** for a **conductor** and end up with something **really small** (e.g. 1×10^{-7} Ω), go back and **check** that you've **converted** your **area** into **m²**.

It's really easy to make mistakes with this equation by leaving the area in **cm²** or **mm²**.

Resistance and Resistivity

For an **Ohmic Conductor**, **R** is a **Constant**

A chap called **Ohm** did most of the early work on resistance. He developed a rule to **predict** how the **current** would **change** as the applied **potential difference increased**, for **certain types** of conductor.

The rule is now called **Ohm's law** and the conductors that **obey** it (mostly metals) are called **ohmic conductors**.

Provided the **temperature** is **constant**, the **current** through an ohmic conductor is **directly proportional** to the **potential difference** across it.

$$R = \frac{V}{I}$$

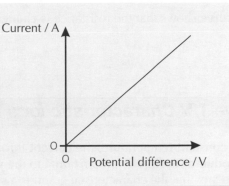

1) As you can see from the graph, **doubling** the **p.d. doubles** the **current**.

2) What this means is that the **resistance** is **constant**.

3) Often **external factors**, such as **light level** or **temperature** will have a **significant effect** on resistance, so you need to remember that Ohm's law is **only** true for **ohmic conductors** at **constant temperature**.

Practice Questions

Q1 Name one environmental factor likely to alter the resistance of a component.

Q2 What is special about an ohmic conductor?

Q3 What three factors does the resistance of a length of wire depend on?

Q4 What are the units for resistivity?

Exam Questions

Q1 Aluminium has a resistivity of $2.8 \times 10^{-8} \, \Omega$ m at 20 °C.

Calculate the resistance of a pure aluminium wire of length 4 m and diameter 1 mm, at 20 °C. [3 marks]

Q2 The table below shows some measurements taken by a student during an experiment investigating an unknown electrical component.

Potential Difference (V)	Current (mA)
2.0	2.67
7.0	9.33
11.0	14.67

(a) Use the first row of the table to calculate the resistance of the component when a p.d. of 2 V is applied. [2 marks]

(b) By means of further calculation, or otherwise, decide whether the component is an ohmic conductor. [3 marks]

Resistance and resistivity are NOT the same...

Superconductors are great. You can use a magnet to set a current flowing in a loop of superconducting wire. Take away the magnet and the current keeps flowing forever. A wire with no resistance never loses any energy. Pretty cool, huh.

I/V Characteristics

Woohoo — real physics. This stuff's actually kind of interesting.

I/V Graphs Show how Resistance Varies

The term '**I/V characteristic**' refers to a **graph** which shows how the **current** (I) flowing through a **component changes** as the **potential difference** (V) across it is increased.

The **shallower** the **gradient** of a characteristic I/V graph, the **greater** the **resistance** of the component.

A **curve** shows that the resistance is **changing**.

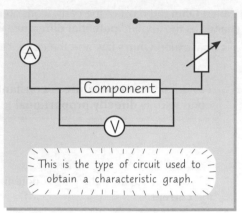

This is the type of circuit used to obtain a characteristic graph.

The I/V Characteristic for a Metallic Conductor is a Straight Line

At **constant temperature**, the **current** through a **metallic conductor** is **directly proportional** to the **voltage**. The fact that the characteristic graph is a **straight line** tells you that the **resistance doesn't change**. **Metallic conductors** are **ohmic** — they have **constant resistance provided** their temperature doesn't change.

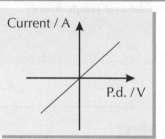

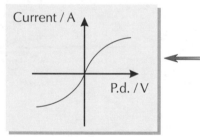

The characteristic graph for a **filament lamp** is a **curve**, which starts **steep** but gets **shallower** as the **voltage rises**. The **filament** in a lamp is just a **coiled up** length of **metal wire**, so you might think it should have the **same characteristic graph** as a **metallic conductor**. It doesn't because it **gets hot**. **Current** flowing through the lamp **increases** its **temperature**.

The **resistance** of a metal **increases** as the **temperature increases**.

The Temperature Affects the Charge Carriers

1) **Charge** is carried through **metals** by **free electrons** in a **lattice** of **positive ions**.
2) Heating up a metal doesn't affect how many electrons there are, but it does make it **harder** for them to **move about**. The **ions vibrate more** when heated, so the electrons **collide** with them more often, **losing energy**.

The **resistance** of most metallic conductors **goes up linearly** with **temperature**.

Semiconductors are Used in Sensors

Semiconductors are **nowhere near** as good at **conducting** electricity as **metals**. This is because there are far, far **fewer charge carriers** available. However, if **energy** is supplied to the semiconductor, **more charge carriers** are often **released**. This means that they make **excellent sensors** for detecting **changes** in their **environment**.

You need to know about **three** semiconductor components — **thermistors**, **LDRs** and **diodes**.

I/V Characteristics

The **Resistance** of a **Thermistor** Depends on **Temperature**

Thermistor circuit symbol: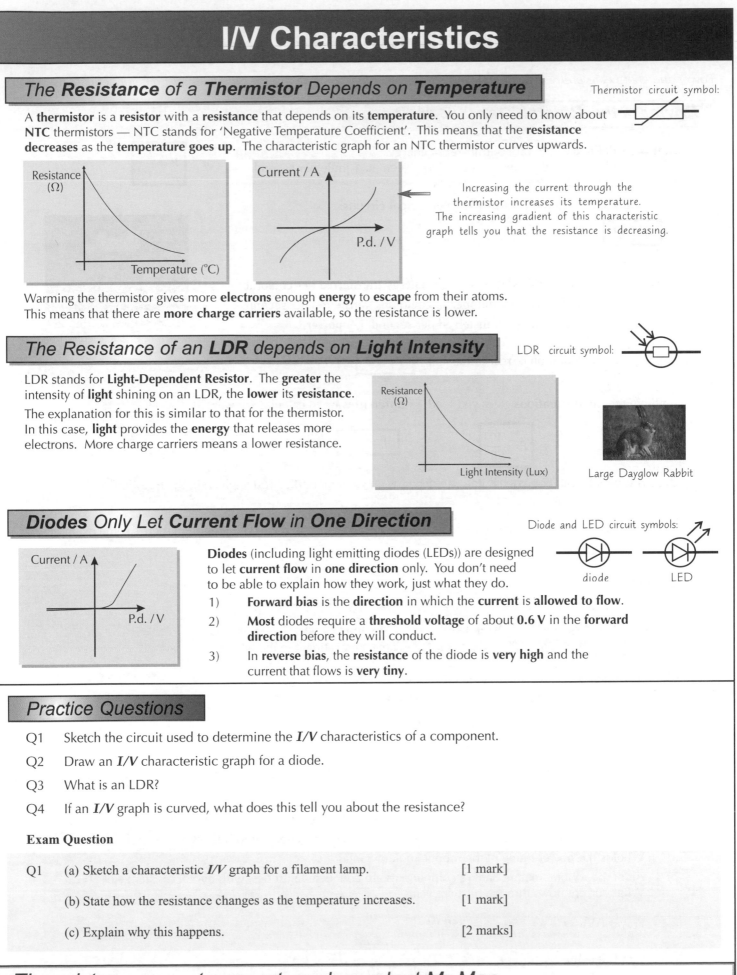

A **thermistor** is a **resistor** with a **resistance** that depends on its **temperature**. You only need to know about **NTC** thermistors — NTC stands for 'Negative Temperature Coefficient'. This means that the **resistance** **decreases** as the **temperature goes up**. The characteristic graph for an NTC thermistor curves upwards.

Increasing the current through the thermistor increases its temperature. The increasing gradient of this characteristic graph tells you that the resistance is decreasing.

Warming the thermistor gives more **electrons** enough **energy** to **escape** from their atoms. This means that there are **more charge carriers** available, so the resistance is lower.

The Resistance of an **LDR** depends on **Light Intensity**

LDR circuit symbol:

LDR stands for **Light-Dependent Resistor**. The **greater** the intensity of **light** shining on an LDR, the **lower** its **resistance**.

The explanation for this is similar to that for the thermistor. In this case, **light** provides the **energy** that releases more electrons. More charge carriers means a lower resistance.

Large Dayglow Rabbit

Diodes Only Let **Current Flow** in **One Direction**

Diode and LED circuit symbols:

diode   LED

Diodes (including light emitting diodes (LEDs)) are designed to let **current flow** in **one direction** only. You don't need to be able to explain how they work, just what they do.

1) **Forward bias** is the **direction** in which the **current** is **allowed to flow**.

2) **Most** diodes require a **threshold voltage** of about **0.6 V** in the **forward direction** before they will conduct.

3) In **reverse bias**, the **resistance** of the diode is **very high** and the current that flows is **very tiny**.

Practice Questions

Q1 Sketch the circuit used to determine the **I/V** characteristics of a component.

Q2 Draw an **I/V** characteristic graph for a diode.

Q3 What is an LDR?

Q4 If an **I/V** graph is curved, what does this tell you about the resistance?

Exam Question

Q1 (a) Sketch a characteristic **I/V** graph for a filament lamp.   [1 mark]

(b) State how the resistance changes as the temperature increases.   [1 mark]

(c) Explain why this happens.   [2 marks]

Thermistor man — temperature-dependent Mr Man...

Learn the graphs on this page, and make sure you can explain them. Whether it's a light-dependent resistor or a thermistor, the same principle applies. More energy releases more charge carriers, and more charge carriers means a lower resistance.

Electrical Energy and Power

Power and energy are pretty familiar concepts — and here they are again. Same principles, just different equations.

Power is the Rate of Transfer of Energy

Power (**P**) is **defined** as the **rate** of **transfer** of **energy**.
It's measured in **watts** (**W**), where **1 watt** is equivalent to **1 joule per second**.

or $P = \dfrac{E}{t}$

There's a really simple formula for **power** in **electrical circuits**:

$$P = VI$$

This makes sense, since:

1) **Potential difference** (**V**) is defined as the **energy transferred** per **coulomb**.

2) **Current** (**I**) is defined as the **number** of **coulombs** transferred per **second**.

3) So **p.d.** × **current** is **energy transferred per second**, i.e. **power**.

He didn't know when, he didn't know where... but one day this PEt would get his revenge.

You know from the definition of **resistance** that: $V = IR$

Combining the **two equations** gives you loads of **different ways** to **calculate power**.

$$P = VI \qquad P = \dfrac{V^2}{R} \qquad P = I^2R$$

Obviously, which equation you should use depends on what **quantities** you're given in the **question**.

Phew... that's quite a few equations to learn and love. And as if they're not exciting enough, here's some examples to get your teeth into...

Example 1

A 24 W car head lamp is connected to a 12 V car battery.

(a) How much energy will the lamp convert into light and heat energy in 2 hours?

(b) Find the total resistance of the lamp and the wires connecting it to the battery.

(a)   Number of seconds in an hour = 120 × 60 = 7200 s

   $E = P \times t = 24 \times 7200 = 172\,800\text{ J} = \underline{\textbf{172.8 kJ}}$

(b)   Rearrange the equation $P = \dfrac{V^2}{R}$, $R = \dfrac{V^2}{P} = \dfrac{12^2}{24} = \dfrac{144}{24} = \underline{\textbf{6 }\Omega}$

Example 2

A robotic mutant Santa from the future converts 750 J of electrical energy into heat every second.

(a)   What is the power rating of the robotic mutant Santa?

(b)   All of the robotic mutant Santa's components are connected in series, with a total resistance of 30 Ω. What current flows through his wire veins?

(a)   Power (W) = $E \div t = 750 \div 1 = \underline{\textbf{750 W}}$

(b)   Rearrange the equation $P = I^2R$, $I = \sqrt{\dfrac{P}{R}} = \sqrt{\dfrac{750}{30}} = \sqrt{25} = \underline{\textbf{5 A}}$

Electrical Energy and Power

Energy *is Easy to* Calculate *if you Know the* Power

Sometimes it's the **total energy** transferred that you're interested in. In this case you simply need to **multiply** the **power** by the **time**. So:

$$E = VIt$$

$\left(\text{or } E = \dfrac{V^2}{R}t \quad \text{or } E = I^2Rt\right)$

You've got to make sure that the time is in seconds.

Example

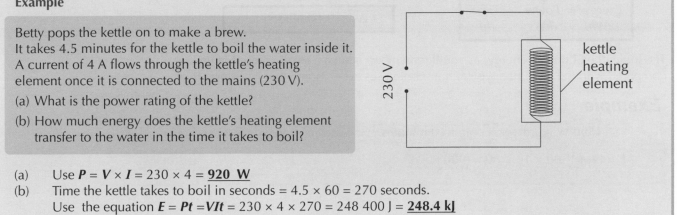

Betty pops the kettle on to make a brew.
It takes 4.5 minutes for the kettle to boil the water inside it.
A current of 4 A flows through the kettle's heating element once it is connected to the mains (230 V).

(a) What is the power rating of the kettle?

(b) How much energy does the kettle's heating element transfer to the water in the time it takes to boil?

230 V

kettle heating element

(a)   Use $P = V \times I = 230 \times 4 = \underline{\textbf{920 W}}$
(b)   Time the kettle takes to boil in seconds = 4.5 × 60 = 270 seconds.
       Use the equation $E = Pt = VIt = 230 \times 4 \times 270 = 248\,400\text{ J} = \underline{\textbf{248.4 kJ}}$

Practice Questions

Q1   Write down the equation linking power, current and resistance.

Q2   What equation links power, voltage and resistance?

Q3   Power is measured in watts. What is 1 watt equivalent to?

Exam Questions

Q1   This question concerns a mains powered hairdryer, the circuit diagram for which is given below.

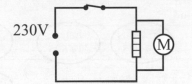

230V

M

(a) The heater has a power of 920 W in normal operation. Calculate the current in the heater.   [2 marks]

(b) The motor has a resistance of 190 Ω. What current will flow in the motor when the hairdryer is used?   [2 marks]

(c) Show that the total power of the hairdryer in normal operation is just under 1.2 kW.   [2 marks]

Q2   A 12 V car battery supplies a current of 48 A for 2 seconds to the car's starter motor.
The total resistance of the connecting wires is 0.01 Ω.

(a) Calculate the energy transferred from the battery.   [1 mark]

(b) Calculate the energy wasted as heat in the wires.   [2 marks]

Ultimate cosmic powers...

Whenever you get equations in this book, you know you're gonna have to learn them. Fact of life.
I used to find it helped to stick big lists of equations all over my walls in the run up to the exams. But as that's
possibly the least cool wallpaper imaginable, I don't advise inviting your friends round till after the exams...

Domestic Energy and Fuses

If you went into an electricity shop and asked for a 100 joule packet of electricity you'd be laughed out of town. Why — because electricity companies use units, not joules — phew, you kids don't know anything these days.

Electricity Companies *don't use* Joules *and* Watts

Electricity companies charge their customers for '**units**' of electricity. Another name for a unit is a **kilowatt-hour (kWh)**. If you know the **power** of an **appliance** and the **length of time** it's used for you can work out the **energy** it uses in kWh.

$$Energy = Power \times Time$$
$$(kWh) \quad (kW) \quad (h)$$

$$1 \text{ kWh} = 3.6 \text{ million joules}$$

1 kW = 1000 W
1 hour = 3600 seconds

The **joule** (the **SI unit** of **energy**, as you'll remember) is such a **small amount** of energy compared with the amount a typical household uses every month that it's **impractical**.

Example

> A **1500 W** hairdryer is on for **10 minutes**. How much energy does it use in J and kWh?

$$E = Pt = 1500 \times 10 \times 60 = \boxed{900\ 000 \text{ J}}$$

$$E = Pt = 1.5 \times 1/6 = \boxed{0.25 \text{ kWh}}$$

Cost of Electricity *is the* Price per Unit *Times the* Number of Units *Used*

Electricity bills can look like they're written in a strange code — but luckily for you, the examples you'll see at AS are easy to understand. Real ones aren't really that bad either — you just need to know **where to look** to find the **important information**. Take a look at the lovely **example** below:

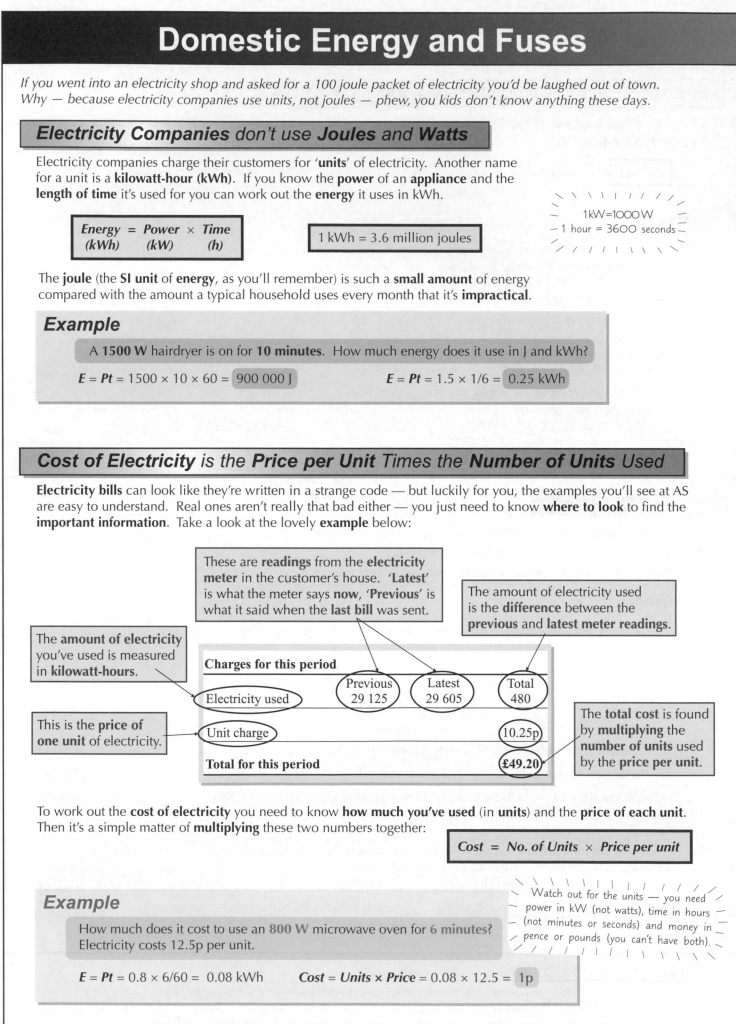

These are **readings** from the **electricity meter** in the customer's house. 'Latest' is what the meter says **now**, 'Previous' is what it said when the **last bill** was sent.

The amount of electricity used is the **difference** between the **previous** and **latest meter readings**.

The **amount of electricity** you've used is measured in **kilowatt-hours**.

This is the **price of one unit** of electricity.

The **total cost** is found by **multiplying** the **number of units** used by the **price per unit**.

Charges for this period

	Previous	Latest	Total
Electricity used	29 125	29 605	480
Unit charge			10.25p
Total for this period			**£49.20**

To work out the **cost of electricity** you need to know **how much you've used** (in **units**) and the **price of each unit**. Then it's a simple matter of **multiplying** these two numbers together:

$$Cost = No. \text{ of Units} \times Price \text{ per unit}$$

Watch out for the units — you need power in kW (not watts), time in hours (not minutes or seconds) and money in pence or pounds (you can't have both).

Example

> How much does it cost to use an **800 W** microwave oven for **6 minutes**? Electricity costs 12.5p per unit.

$$E = Pt = 0.8 \times 6/60 = \boxed{0.08 \text{ kWh}} \qquad Cost = Units \times Price = 0.08 \times 12.5 = \boxed{1p}$$

Domestic Energy and Fuses

Fuses Prevent Shocks and Fire

A **fuse** is a very **fine wire** in a glass tube that's connected between the **live terminal** of the mains supply and an **appliance**. If the **current** in the circuit gets too **big** (bigger than the fuse rating — see below), the fuse wire **melts** and **breaks** the circuit. Fuses should be rated as near as possible but **just higher** than the normal operating current.

The **earth wire** and **fuse** in an appliance with a metal case work together like this:

1) The earth pin in the plug is connected to the **case** of the appliance via the **earth wire**.

2) A **fault** can develop in which the **live** somehow **touches** the case. Then because the case is earthed, a big current flows **in** through the **live**, through the **case** and **out** down the **earth** wire.

3) This **surge** in current blows the fuse, which cuts off the live supply. This prevents electric shocks from the case.

You Can Work Out What Fuse to Use from the Appliance's Power Rating

Most electrical goods have a plate showing their **power rating** and **voltage rating**. To work out the fuse needed, you have to work out the current that the appliance will normally draw.

Example A toaster is rated at 2.2 kW, 230 V. Find the fuse needed.

You can usually only get fuses with ratings of 3 A, 5 A or 13 A.

Use $P = VI$, and rearrange to give $I = P / V = 2200 / 230 = $ **9.57 A**.
The fuse should be rated just a bit higher than the normal current, so this toaster should have a 13 A fuse.

Practice Questions

Q1 Why aren't joules used on electricity bills? What is used instead?
Q2 What equation links power, energy and time?
Q3 What equation would you use to find the cost of electricity?
Q4 Describe how a fuse works.

Exam Questions

Q1 A vacuum cleaner is rated at 1800 W.

(a) Calculate the energy transferred when the vacuum cleaner is operated for 15 minutes.
Give your answer in: (i) joules, (ii) kilowatt-hours. [2 marks]

(b) Calculate the cost of using the vacuum cleaner for 15 minutes.
Electricity costs 14.6 pence per unit. [1 mark]

Q2 A television is rated at 1500 W, 230 V. Electricity costs 9.8 pence per unit.

(a) Find the fuse needed for the television. [2 marks]

(b) What is the cost of using the television for two and a quarter hours? [2 marks]

When the television is in standby mode, it draws a current of 6.5 mA.

(c) Show that the power of the television in standby mode is approximately 1.5 W. [1 mark]

(d) Estimate the cost of leaving the television on standby overnight (10 hours). [2 marks]

Hurrah — now my toaster won't kill me...

Aren't fuses wonderful things? They're much better than electricity bills, that's for sure. Now, I know there are a couple of equations on these two pages, but you probably met them at GCSE — and they're fairly straightforward, so don't panic. Instead, try the practice exam questions — go on, I know you'll like them — one of them's even about television.

E.m.f. and Internal Resistance

There's resistance everywhere — inside batteries, in all the wires and in the components themselves.
No one's for giving current an easy ride.

Batteries have Resistance

From now on, I'm assuming that the resistance of the wires in the circuit is zero. In practice, they do have a small resistance.

Resistance comes from **electrons colliding** with **atoms** and **losing energy**.

In a **battery**, **chemical energy** is used to make **electrons move**. As they move, they collide with atoms inside the battery — so batteries **must** have resistance. This is called **internal resistance**.

Internal resistance is what makes **batteries** and **cells warm up** when they're used.

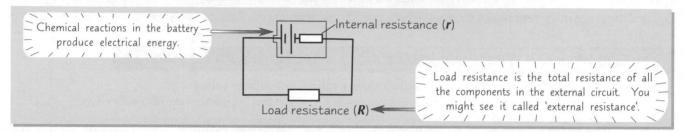

Chemical reactions in the battery produce electrical energy.

Internal resistance (**r**)

Load resistance is the total resistance of all the components in the external circuit. You might see it called 'external resistance'.

Load resistance (**R**)

1) The amount of **electrical energy** the battery produces for each **coulomb** of charge is called its **electromotive force** or **e.m.f.** (ε). Be careful — e.m.f. **isn't** actually a force. It's measured in **volts**.

2) The **potential difference** across the **load resistance** (**R**) is the **energy transferred** when **one coulomb** of charge flows through the **load resistance**. This potential difference is called the **terminal p.d.** (**V**).

3) If there was **no internal resistance**, the **terminal p.d.** would be the **same** as the **e.m.f.** However, in **real** power supplies, there's **always some energy lost** overcoming the internal resistance.

4) The **energy wasted per coulomb** overcoming the internal resistance is called the **lost volts** (**v**).

Conservation of energy tells us:

energy per coulomb supplied by the source	=	energy per coulomb used in load resistance	+	energy per coulomb wasted in internal resistance

There are Loads of Calculations with E.m.f. and Internal Resistance

Examiners can ask you to do **calculations** with **e.m.f.** and **internal resistance** in loads of **different** ways. You've got to be ready for whatever they throw at you.

$$\varepsilon = V + v \qquad \varepsilon = I(R + r)$$
$$V = \varepsilon - v \qquad \varepsilon = V + Ir$$

Learn these equations for the exam. Only this one will be on your formula sheet.

These are all basically the **same equation**, just written differently. If you're given enough information you can calculate the e.m.f. (ε), terminal p.d. (**V**), lost volts (**v**), current (**I**), load resistance (**R**) or internal resistance (**r**). Which equation you should use depends on what information you've got, and what you need to calculate.

Most Power Supplies Need Low Internal Resistance

A **car battery** has to deliver a **really high current** — so it needs to have a **low internal resistance**. The cells used to power a **torch** or a **personal stereo** are the same. **Generally**, batteries have an **internal resistance** of **less than 1 Ω**.

Since **internal resistance** causes **energy loss**, you'd think **all** power supplies should have a **low internal resistance**.

High voltage power supplies are the **exception**. **HT** (high tension) and **EHT** (extremely high tension) **supplies** are designed with **very high** internal resistances. This means that if they're **accidentally short-circuited** only a **very small current** can flow. Much **safer**.

E.m.f. and Internal Resistance

Use this **Circuit** to **Measure Internal Resistance** and **E.m.f.**

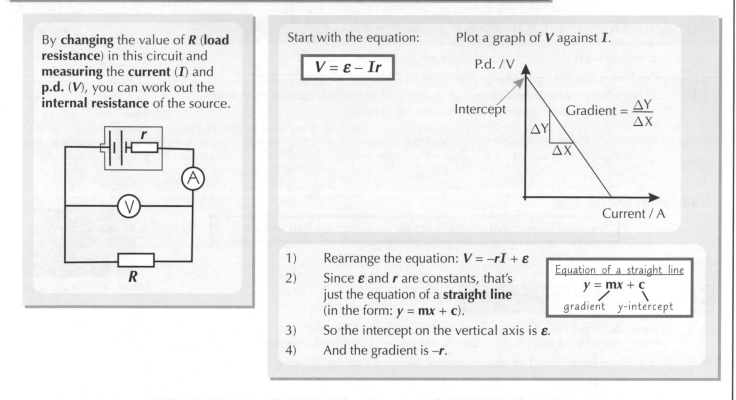

By **changing** the value of **R** (**load resistance**) in this circuit and **measuring** the **current** (**I**) and **p.d.** (**V**), you can work out the **internal resistance** of the source.

Start with the equation:

$$V = \varepsilon - Ir$$

Plot a graph of **V** against **I**.

1) Rearrange the equation: $V = -rI + \varepsilon$
2) Since ε and **r** are constants, that's just the equation of a **straight line** (in the form: $y = mx + c$).
3) So the intercept on the vertical axis is ε.
4) And the gradient is $-r$.

Equation of a straight line
$y = mx + c$
gradient   y-intercept

An **easier** way to **measure** the **e.m.f.** of a **power source** is by connecting a high-resistance **voltmeter** across its **terminals**. A **small current flows** through the **voltmeter**, so there must be some **lost volts** — this means you measure a value **very slightly less** than the **e.m.f.** In **practice** the difference **isn't** usually **significant**.

Practice Questions

Q1 What causes internal resistance?

Q2 What is meant by 'lost volts'?

Q3 What is the difference between e.m.f. and terminal p.d.?

Q4 Write the equation used to calculate the terminal p.d. of a power supply.

Exam Questions

Q1 A large battery with an internal resistance of 0.8 Ω and e.m.f. 24 V is used to power a dentist's drill with resistance 4 Ω.

(a) Calculate the current in the circuit when the drill is connected to the power supply. [2 marks]

(b) Calculate the voltage across the drill while it is being used. [1 mark]

Q2 A student mistakenly connects a 10 Ω ray box to an HT power supply of 500 V.
The ray box does not light, and the student measures the current flowing to be only 50 mA.

(a) Calculate the internal resistance of the HT power supply. [2 marks]

(b) Explain why this is a sensible internal resistance for an HT power supply. [2 marks]

You're UNBELIEVABLE... [Frantic air guitar]... Ueuuurrrrghhh... Yeah...

Wanting power supplies to have a low internal resistance makes sense — you wouldn't want your MP3 player battery melting if you listened to music for more than half an hour. Make sure you know your e.m.f. equations — they're an exam fave. A good way to get them learnt is to keep trying to get from one equation to another... dull, but it can help.

Conservation of Energy & Charge in Circuits

There are some things in Physics that are so fundamental that you just have to accept them. Like the fact that there's loads of Maths in it. And that energy is conserved. And that Physicists get more homework than everyone else.

Charge Doesn't 'Leak Away' Anywhere — it's Conserved

1) As **charge flows** through a circuit, it **doesn't** get **used up** or **lost**.

2) This means that whatever **charge flows into** a junction will **flow out** again.

3) Since **current** is **rate of flow of charge**, it follows that whatever **current flows into** a junction is the same as the current **flowing out** of it.

e.g.

$$\text{CHARGE FLOWING IN 1 SECOND}$$
$$Q_1 = 6\,\text{C} \Rightarrow I_1 = 6\,\text{A}$$
$$Q_2 = 2\,\text{C} \Rightarrow I_2 = 2\,\text{A}$$
$$Q_3 = 4\,\text{C} \Rightarrow I_3 = 4\,\text{A}$$
$$I_1 = I_2 + I_3$$

Kirchhoff's first law says:

> The total **current entering a junction** = the total **current leaving it.**

Energy conservation is vital.

Energy is Conserved too

1) **Energy is conserved.** You already know that. In **electrical circuits**, **energy** is **transferred round** the circuit. Energy **transferred to** a charge is **e.m.f.**, and energy **transferred from** a charge is **potential difference**.

2) In a **closed loop**, these two quantities must be **equal** if energy is conserved (which it is).

Kirchhoff's second law says:

> The **total e.m.f.** around a **series circuit** = the **sum** of the **p.d.s** across each component. (or $\varepsilon = \Sigma IR$ in symbols)

Exam Questions get you to Apply **Kirchhoff's Laws** to Combinations of **Resistors**

A **typical exam question** will give you a **circuit** with bits of information missing, leaving you to fill in the gaps. Not the most fun... but on the plus side you get to ignore any internal resistance stuff (unless the question tells you otherwise)... hurrah. You need to remember the **following rules**:

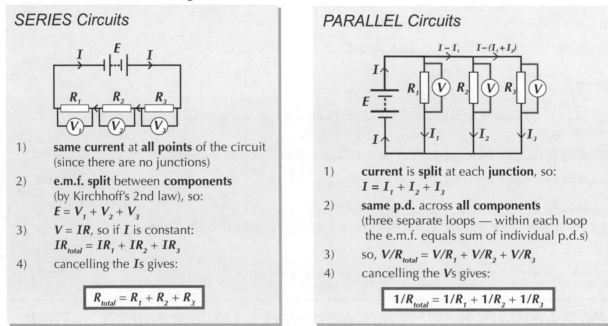

SERIES Circuits

1) **same current** at **all points** of the circuit (since there are no junctions)

2) **e.m.f. split** between **components** (by Kirchhoff's 2nd law), so:
$$E = V_1 + V_2 + V_3$$

3) $V = IR$, so if I is constant:
$$IR_{total} = IR_1 + IR_2 + IR_3$$

4) cancelling the Is gives:

$$R_{total} = R_1 + R_2 + R_3$$

PARALLEL Circuits

1) **current** is **split** at each **junction**, so:
$$I = I_1 + I_2 + I_3$$

2) **same p.d.** across **all components** (three separate loops — within each loop the e.m.f. equals sum of individual p.d.s)

3) so, $V/R_{total} = V/R_1 + V/R_2 + V/R_3$

4) cancelling the Vs gives:

$$1/R_{total} = 1/R_1 + 1/R_2 + 1/R_3$$

...and there's an example on the next page to make sure you know what to do with all that...

Conservation of Energy & Charge in Circuits

Worked Exam Question

A battery of e.m.f. 16 V and negligible internal resistance is connected in a circuit as shown:

a) Show that the group of resistors between X and Y could be replaced by a single resistor of resistance 15 Ω.

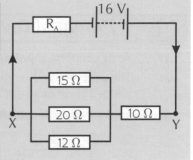

You can find the **combined resistance** of the 15 Ω, 20 Ω and 12 Ω resistors using:

$1/R = 1/R_1 + 1/R_2 + 1/R_3 = 1/15 + 1/20 + 1/12 = 1/5$ $\Rightarrow R = 5\ \Omega$

So **overall resistance** between **X** and **Y** can be found by $R = R_1 + R_2 = 5 + 10 = \mathbf{15\ \Omega}$

b) If $R_A = 20\ \Omega$:
 (i) calculate the potential difference across R_A,

Careful — there are a few steps here. You need the p.d. across R_A, but you don't know the current through it. So start there:

total resistance in circuit = 20 + 15 = 35 Ω, **so** current through R_A can be found using $I = V_{total}/R_{total}$:

$I = 16/35$ A

then you can use $V = IR_A$ to find the p.d. across R_A: $V = 16/35 \times 20 = \mathbf{9.1\ V}$

 (ii) calculate the current in the 15 Ω resistor.

You know the **current flowing** into the group of three resistors and out of it, but not through the individual branches. But you know that their **combined resistance** is **5 Ω** (from part a) so you can work out the p.d. across the group:

$V = IR = 16/35 \times 5 = 16/7$ V

The p.d. across the **whole group** is the same as the p.d. across each **individual resistor**, so you can use this to find the current through the 15 Ω resistor:

$I = V/R = (16/7) / 15 = \mathbf{0.15\ A}$

Practice Questions

Q1 State Kirchhoff's laws.

Q2 Find the current through and potential difference across each of two 5 Ω resistors when they are placed in a circuit containing a 5 V battery, and are wired: a) in series, b) in parallel.

Exam Question

Q1 For the circuit on the right:

(a) Calculate the total effective resistance of the three resistors in this combination. [2 marks]

(b) Calculate the main current, I_3. [2 marks]

(c) Calculate the potential difference across the 4 Ω resistor. [1 mark]

(d) Calculate the potential difference across the parallel pair of resistors. [1 mark]

(e) Using your answer from 1 (d), calculate the currents I_1 and I_2. [2 marks]

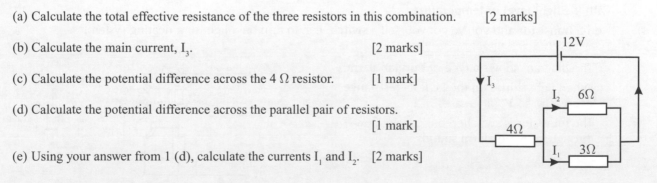

This is a very purple page — needs a bit of yellow I think...

V = IR is the formula you'll use most often in these questions. Make sure you know whether you're using it on the overall circuit, or just one specific component. It's amazingly easy to get muddled up — you've been warned.

The Potential Divider

I remember the days when potential dividers were pretty much the hardest thing they could throw at you. Then along came AS Physics. Hey ho.

Anyway, in context this doesn't seem too hard now, so get stuck in.

Use a **Potential Divider** to get a **Fraction** of a **Source Voltage**

1) At its simplest, a **potential divider** is a circuit with a **voltage source** and a couple of **resistors** in series.

2) The **potential** of the voltage source (e.g. a power supply) is **divided** in the **ratio** of the **resistances**. So, if you had a **2 Ω** resistor and a **3 Ω** resistor, you'd get **2/5** of the p.d. across the **2 Ω** resistor and **3/5** across the **3 Ω**.

3) That means you can **choose** the **resistances** to get the **voltage** you **want** across one of them.

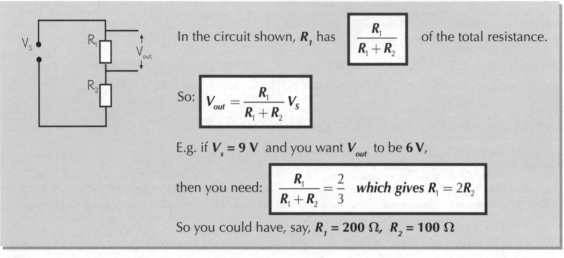

In the circuit shown, R_1 has $\dfrac{R_1}{R_1 + R_2}$ of the total resistance.

So: $$V_{out} = \frac{R_1}{R_1 + R_2} V_s$$

E.g. if $V_s = 9\,V$ and you want V_{out} to be **6 V**,

then you need: $\dfrac{R_1}{R_1 + R_2} = \dfrac{2}{3}$ *which gives* $R_1 = 2R_2$

So you could have, say, $R_1 = 200\,\Omega$, $R_2 = 100\,\Omega$

4) This circuit is mainly used for **calibrating voltmeters**, which have a **very high resistance**.

5) If you put something with a **relatively low resistance** across R_1 though, you start to run into **problems**. You've **effectively** got **two resistors** in **parallel**, which will **always** have a **total** resistance **less** than R_1. That means that V_{out} will be **less** than you've calculated, and will depend on what's connected across R_1. Hrrumph.

Add an **LDR** or **Thermistor** for a **Light** or **Temperature Switch**

1) A **light-dependent resistor** (LDR) has a very **high resistance** in the **dark**, but a **lower resistance** in the **light**.

2) An **NTC thermistor** has a **high resistance** at **low temperatures**, but a much **lower resistance** at **high temperatures** (it varies in the opposite way to a normal resistor, only much more so).

3) Either of these can be used as one of the **resistors** in a **potential divider**, giving an **output voltage** that **varies** with the **light level** or **temperature**.

4) Add a **transistor** and you've got yourself a **switch**, e.g. to turn on a light or a heating system.

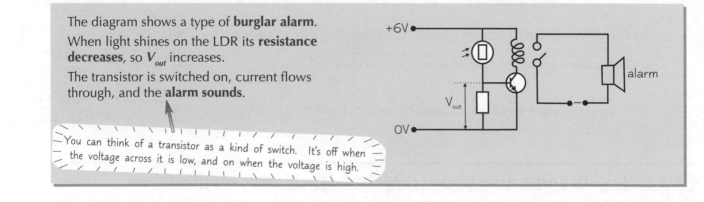

The diagram shows a type of **burglar alarm**.

When light shines on the LDR its **resistance decreases**, so V_{out} increases.

The transistor is switched on, current flows through, and the **alarm sounds**.

You can think of a transistor as a kind of switch. It's off when the voltage across it is low, and on when the voltage is high.

The Potential Divider

A *Potentiometer* uses a *Variable Resistor* to give a *Variable Voltage*

1) A **potentiometer** has a variable resistor replacing R_1 and R_2 of the potential divider, but it uses the **same idea** (it's even sometimes **called** a potential divider just to confuse things).

2) You move a **slider** or turn a knob to **adjust** the **relative sizes** of R_1 and R_2. That way you can vary V_{out} from **0 V** up to the source voltage.

3) This is dead handy when you want to be able to **change** a **voltage continuously**, like in the **volume control** of a stereo.

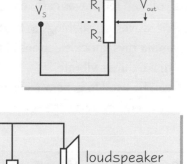

Here, V_s is replaced by the input signal (e.g. from a CD player) and V_{out} is the output to the amplifier and loudspeaker.

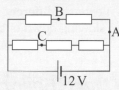

Practice Questions

Q1 Look at the burglar alarm circuit on page 56. How could you change the circuit so that the alarm sounds when the light level decreases?

Q2 The LDR in the burglar alarm circuit has a resistance of 300 Ω when light and 900 Ω when dark. The fixed resistor has a value of 100 Ω. Show that V_{out} (light) = 1.5 V and V_{out} (dark) = 0.6 V.

Exam Questions

Q1 In the circuit on the right, all the resistors have the same value. Calculate the p.d. between:

(i) A and B. [1 mark]

(ii) A and C. [1 mark]

(iii) B and C. [1 mark]

Q2 Look at the circuit on the right.

(a) Calculate the p.d. between A and B as shown by a high resistance voltmeter placed between the two points. [1 mark]

(b) A 40 Ω resistor is now placed between points A and B. Calculate the p.d. across AB and the current flowing through the 40 Ω resistor. [4 marks]

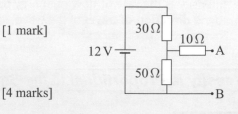

OI...YOU... [bang bang bang]... turn that potentiometer down...

You'll probably have to use a potentiometer in every experiment you do with electricity from now on in, so you'd better get used to them. I can't stand the things myself, but then lab and me don't mix — far too technical.

The Nature of Waves

Aaaah... playing with slinky springs and waggling ropes about. It's all good clean fun as my mate Richard used to say...

A **Wave Transfers Energy** Away from Its Source

A **progressive** (moving) wave carries **energy** from one place to another **without transferring any material**. Here are some ways you can tell waves carry energy:

1) Electromagnetic waves cause things to **heat up**.

2) **X-rays** and **gamma rays** knock electrons out of their orbits, causing **ionisation**.

3) Loud **sounds** make things **vibrate**.

4) **Wave power** can be used to **generate electricity**.

5) Since waves carry energy away, the **source** of the wave **loses energy**.

Here are all the **bits** of a **Wave** you Need to Know

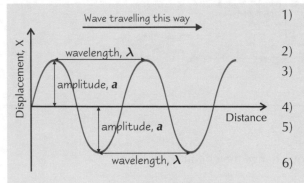

1) **Displacement**, *X*, metres — how far a **point** on the wave has **moved** from its **undisturbed position**.

2) **Amplitude**, a, metres — **maximum displacement**.

3) **Wavelength**, λ, metres — the **length** of **one whole wave**, from **crest** to **crest** or **trough** to **trough**.

4) **Period**, *T*, seconds — the **time taken** for a **whole vibration**.

5) **Frequency**, *f*, hertz — the **number** of **vibrations per second** passing a given **point**.

6) **Phase difference** — the amount by which **one wave lags behind another** wave. **Measured** in **degrees** or **radians**. See page 64.

Waves Can Be **Reflected** and **Refracted**

Reflection — the wave is **bounced back** when it **hits a boundary**. E.g. you can see the reflection of light in mirrors. The reflection of water waves can be demonstrated in a ripple tank.

Refraction — the wave **changes direction** as it enters a **different medium**. The change in direction is a result of the wave slowing down or speeding up.

Intensity is a Measure of **How Much Energy** a Wave is Carrying

1) When you talk about "**brightness**" for light or "**loudness**" for sound, what you really mean is **how much light** or **sound** energy hits your eyes or your ears **per second**.

2) The scientific measure of this is **intensity**.

> Intensity is the **rate of flow** of energy per **unit area** at **right angles** to the **direction of travel** of the wave. It's measured in **Wm^{-2}**.

Intensity is **Proportional** to the **Square** of the **Amplitude** of the **Wave**

$$I \propto A^2$$

1) This comes from the fact that intensity is proportional to energy, and the energy of a wave depends on the square of the amplitude.

2) From this you can tell that for a vibrating source it takes four times as much energy to double the size of the vibrations.

The Nature of Waves

The **Frequency** is the **Inverse** of the **Period**

$$\text{Frequency} = \frac{1}{\text{period}}$$

It's that simple.
Get the **units** straight: **1 Hz = 1 s⁻¹**.

Wave Speed, Frequency and Wavelength are Linked by the Wave Equation

Wave speed can be measured just like the speed of anything else:

$$\text{Speed } (v) = \frac{\text{distance moved } (d)}{\text{time taken } (t)}$$

Remember, you're not measuring how fast a physical point (like one molecule of rope) moves. You're measuring how fast a point on the **wave pattern** moves.

Learn the **Wave Equation**...

Speed of wave (v) = wavelength (λ) × frequency (f)

$$v = \lambda f$$

You need to be able to rearrange this equation for v, λ or f.

... and How to **Derive** it

You can work out the **wave equation** by imagining **how long** it takes for the **crest** of a wave to **move** across a **distance** of **one wavelength**. The **distance travelled** is λ. **By definition**, the **time taken** to travel **one whole wavelength** is the **period** of the wave, which is equal to **1/f**.

$$\text{Speed } (v) = \frac{\text{distance moved } (d)}{\text{time taken } (t)} \quad \longrightarrow \quad \text{Speed } (v) = \frac{\text{distance moved } (\lambda)}{\text{time taken } (1/f)}$$

Learn to recognise when to use **$v = \lambda f$** and when to use **$v = d/t$**. Look at which variables are mentioned in the question.

Practice Questions

Q1　Does a wave carry matter **or** energy from one place to another?

Q2　Diffraction and interference are two wave properties. Write down two more.

Q3　Write down the relationship between the amplitude of a wave and its intensity.

Q4　Give the units of frequency, displacement and amplitude.

Q5　Write down the equation connecting v, λ and f.

Exam Question

Q1　A buoy floating on the sea takes 6 seconds to rise and fall once (complete a full period of oscillation). The difference in height between the buoy at its lowest and highest points is 1.2 m, and waves pass it at a speed of 3 ms⁻¹.

(a) How long are the waves?　[2 marks]

(b) What is the amplitude of the waves?　[1 mark]

Learn the wave equation and its derivation — pure poetry...

This isn't too difficult to start you off — most of it you'll have done at GCSE anyway. But once again, it's a whole bunch of equations to learn, and you won't get far without learning them. Yada yada.

Longitudinal and Transverse Waves

There are different types of wave — and the difference is easiest to see using a slinky. Try it — you'll have hours of fun.

In **Transverse Waves** the **Vibration** is at **Right Angles** to the **Direction** of Travel

All **electromagnetic waves** are **transverse**. Other examples of transverse waves are **ripples** on water and waves on **ropes**.

There are **two** main ways of **drawing** transverse waves:

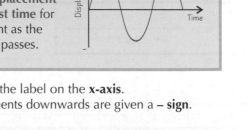

Vibrations from side to side

Wave travelling this way

① They can be shown as **graphs** of **displacement** against **distance** along the path of the wave.

Displacement

crest

λ

Distance

trough

② Or, they can be shown as graphs of **displacement against time** for a point as the wave passes.

Displacement

Time

Both sorts of graph often give the **same shape**, so make sure you check out the label on the **x-axis**.
Displacements **upwards** from the centre line are given a **+ sign**. Displacements downwards are given a **– sign**.

In **Longitudinal Waves** the **Vibrations** are **Along** the Direction of Travel

The most **common** example of a **longitudinal wave** is **sound**. A sound wave consists of alternate **compressions** and **rarefactions** of the **medium** it's travelling through. (That's why sound can't go through a vacuum.) Some types of **earthquake shock waves** are also longitudinal.

Compression   Rarefaction

Vibrations in same direction as wave is travelling

One wavelength

It's hard to **represent** longitudinal waves **graphically**. You'll usually see them plotted as **displacement** against **time**. These can be **confusing** though, because they look like a **transverse wave**.

A **Polarised Wave** only **Oscillates** In One Direction

1) If you **shake a rope** to make a **wave** you can move your hand **up and down** or **side to side** or in a **mixture** of directions — it still makes a **transverse wave**.

2) But if you try to pass **waves in a rope** through a **vertical fence**, the wave will only get through if the **vibrations** are **vertical**. The fence filters out vibration in other directions. This is called **polarising** the wave.

3) Ordinary **light waves** are a mixture of **different directions** of **vibration**. (The things vibrating are electric and magnetic fields.) A **polarising filter** only transmits vibrations in one direction.

4) If you have two polarising filters at **right angles** to each other, then **no** light will get through.

5) Polarisation **can only happen** for **transverse** waves. The fact that you can polarise light is one **proof** that it's a transverse wave.

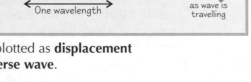

direction of waves

rope

fence

When **Light Reflects** it is **Partially Polarised**

1) Rotating a **polarising filter** in a beam of light shows the fraction of the light that is vibrating in each **direction**.

2) If you direct a beam of unpolarised light at a reflective surface then view the **reflected ray** through a polarising filter, the intensity of light leaving the filter **changes** with the **orientation** of the filter.

3) The intensity changes because light is **partially polarised** when it is **reflected**.

4) This effect is used to remove **unwanted reflections** in photography and in **Polaroid sunglasses** to remove **glare**.

Unpolarised light

Glass block

Partially polarised light

When the light reaches the glass block, it is reflected and polarised.

As the polarising filter is rotated, the intensity of light leaving it changes.

Longitudinal and Transverse Waves

Materials Can Rotate the Plane of Polarisation

The **plane** in which a wave moves and **vibrates** is called the **plane of polarisation** — e.g. the rope on the last page was polarised in the **vertical** plane by the fence. Some **materials** (e.g. crystals) **rotate** the plane of polarised light. You can **measure** how much a material rotates the plane of polarised light using two **polarising filters**:

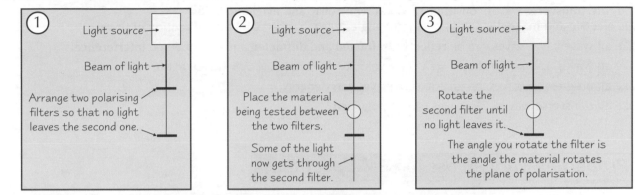

① Light source → Beam of light → Arrange two polarising filters so that no light leaves the second one.

② Light source → Beam of light → Place the material being tested between the two filters. Some of the light now gets through the second filter.

③ Light source → Beam of light → Rotate the second filter until no light leaves it. The angle you rotate the filter is the angle the material rotates the plane of polarisation.

Rotating the Plane of Polarisation Affects the Intensity

You've seen that passing light through a polarising filter **rotates** its **plane of polarisation**, but it also changes the **amplitude** and **intensity** of the transmitted wave.

The **amplitude** of the **transmitted** wave is the **component** of the **incident** wave in the direction of the **new plane** of polarisation.

$$A = A_0 \cos \theta$$

Where A is the amplitude of the **transmitted** wave, A_0 is the amplitude of the **incident** wave and θ (theta) is the **angle** the plane has been rotated.

The **intensity** of the **transmitted** light is **proportional** to the **amplitude squared** — this is **Malus' law**.

$$I = I_0 \cos^2 \theta$$

Where I is the intensity of transmitted light, I_0 is the intensity of incident light and θ (theta) is the **angle** the plane has been rotated.

Practice Questions

Q1 Give examples of a transverse wave and a longitudinal wave.

Q2 What is a polarised wave? How can you polarise a wave?

Q3 What is Malus' law and what is it used for?

Exam Questions

Q1 In an experiment, light is shone through a disc of a crystal called "Iceland spar". The beam of light is less bright when it emerges from the crystal than when it enters. Next, a second identical disc of Iceland spar is placed in front of the first. The first disc is held steady while the second is rotated (in the plane of the disc). The intensity of light emerging changes as the second disc rotates. At two points in each rotation, no light gets through at all.

(a) Explain the results of these experiments. You may use a diagram to help your answer. [5 marks]

(b) When the second disc is rotated to angle α, the intensity of light emerging from the second disc is half the value it was when it left the first disc. Calculate angle α. [3 marks]

Q2 Give one example of an application of polarisation and explain how it works. [2 marks]

Caution — rotating the plane may cause nausea...

The waves broadcast from TV or radio transmitters are polarised. So you have to line up the receiving aerial with the transmitting aerial to receive the signal properly. It's one reason why the TV picture's lousy if the aerial gets knocked.

The Electromagnetic Spectrum

There's nothing really deep and meaningful to understand on this page — just a load of facts to learn I'm afraid.

All **Electromagnetic Waves** Have Some **Properties** In Common

1) They travel in a **vacuum** at a **speed** of 2.998×10^8 **ms^{-1}**, and at slower speeds in other media.

2) They are **transverse** waves consisting of **vibrating electric** and **magnetic fields**.
The **electric** and **magnetic** fields are at **right angles** to each other and to the **direction of travel**.

3) Like all waves, EM waves can be **reflected**, **refracted** and **diffracted** and can undergo **interference**.

4) Like all waves, EM waves obey $v = f\lambda$ (v = velocity, f = frequency, λ = wavelength).

5) Like all progressive waves, progressive EM waves **carry energy**.

6) Like all transverse waves, EM waves can be **polarised**.

Some **Properties Vary** Across the **EM Spectrum**

EM waves with different wavelengths behave differently in some respects. The spectrum is split into seven categories:
radio waves, **microwaves**, **infrared**, **visible light**, **ultraviolet**, **X-rays** and **gamma rays**.

1) The longer the wavelength, the more **obvious** the wave characteristics — e.g., long radio waves diffract round hills.

2) **Energy** is directly proportional to **frequency**. **Gamma rays** have the **highest energy**; **radio waves** the **lowest**.

3) The **higher** the **energy**, in general the more **dangerous** the wave.

4) The **lower the energy** of an EM wave, the **further from the nucleus** it comes from. **Gamma radiation** comes from
inside the **nucleus**. **X-rays to visible light** come from energy-level transitions in **atoms** (see p. 78). **Infrared**
radiation and **microwaves** are associated with **molecules**. **Radio waves** come from oscillations in **electric fields**.

The **Properties** of an **EM Wave** Change with **Wavelength**

Type	Approximate wavelength / m	Penetration	Uses
Radio waves	$10^{-1} - 10^6$	Pass through matter.	Radio transmissions.
Microwaves	$10^{-3} - 10^{-1}$	Mostly pass through matter, but cause some heating.	Radar. Microwave cookery. TV transmissions.
Infrared (IR)	$7 \times 10^{-7} - 10^{-3}$	Mostly absorbed by matter, causing it to heat up.	Heat detectors. Night-vision cameras. Remote controls. Optical fibres.
Visible light	$4 \times 10^{-7} - 7 \times 10^{-7}$	Absorbed by matter, causing some heating effect.	Human sight. Optical fibres.
Ultraviolet (UV)	$10^{-8} - 4 \times 10^{-7}$	Absorbed by matter. Slight ionisation.	Sunbeds. Security markings that show up in UV light.
X-rays	$10^{-13} - 10^{-8}$	Mostly pass through matter, but cause ionisation as they pass.	To see damage to bones and teeth. Airport security scanners. To kill cancer cells.
Gamma rays	$10^{-16} - 10^{-10}$	Mostly pass through matter, but cause ionisation as they pass.	Irradiation of food. Sterilisation of medical instruments. To kill cancer cells.

The Electromagnetic Spectrum

Different Types of EM Wave Have Different **Effects** on the **Body**

Type	Production	Effect on human body
Radio waves	Oscillating electrons in an aerial	No effect.
Microwaves	Electron tube oscillators. Masers.	Absorbed by water — danger of cooking human body*.
Infrared (IR)	Natural and artificial heat sources.	Heating. Excess heat can harm the body's systems.
Visible light	Natural and artificial light sources.	Used for sight. Too bright a light can damage eyes.
Ultraviolet (UV)	e.g. the Sun.	Tans the skin. Can cause skin cancer and eye damage.
X-rays	Bombarding metal with electrons.	Cancer due to cell damage. Eye damage.
Gamma rays	Radioactive decay of the nucleus.	Cancer due to cell damage. Eye damage.

1) UV radiation is split into categories based on frequency — **UV-A**, **UV-B** and **UV-C**.

2) **UV-A** has the **lowest** frequency and is the **least damaging**, although it's thought to be a significant cause of **skin aging**.

3) Higher-frequency **UV-B** is more dangerous. It can be **absorbed** by DNA molecules, causing **mutations** which can lead to **cancer**. UV-B is responsible for **sunburn** too.

4) **UV-C** has a high enough frequency to be **ionising** — it carries enough energy to knock electrons off atoms. This can cause **cell mutation** or **destruction**, and **cancer**. It's almost **entirely blocked** by the ozone layer, though.

5) **Dark** skin gives some protection from UV rays, stopping them reaching more vulnerable tissues below. So **tanning** is a protection mechanism — **UV-A** triggers the release of melanin (a brown pigment) in the skin.

6) **Sunscreens** provide some protection from UV in sunlight. The **Sun Protection Factor** (**SPF**) of the sunscreen tells you how well it protects against **UV-B** radiation. It **doesn't** tell you anything about the UV-A protection though. Many modern sunscreens include tiny particles of **zinc oxide** and **titanium dioxide** to block UV-A.

* Or small animals.

Practice Questions

Q1 What are the main practical uses of infrared radiation?

Q2 Which types of electromagnetic radiation have the highest and lowest energies?

Q3 What is the significance of the speed 2.998×10^8 ms^{-1}?

Q4 Why are microwaves dangerous?

Q5 How does the energy of an EM wave vary with frequency?

Exam Questions

Q1 In a vacuum, do X-rays travel faster, slower or at the same speed as visible light? Explain your answer. [2 marks]

Q2 (a) Describe briefly the physics behind a practical use of X rays. [2 marks]

(b) What is the difference between gamma rays and X-rays? [2 marks]

Q3 Give an example of a type of electromagnetic wave causing a hazard to health. [2 marks]

I've got UV hair...

No really I have. It's great. It's purple. And it's got shiny glittery white bits in it.
Aaaanyway... moving swiftly on. Loads of facts to learn on these pages. You probably know most of this from GCSE anyway, but make sure you know it well enough to answer a question on it in the exam. Not much fun, but... there you go.

Superposition and Coherence

When two waves get together, it can be either really impressive or really disappointing.

Superposition *Happens When* Two *or* More *Waves* Pass Through *Each Other*

1) At the **instant** the waves **cross**, the **displacements** due to each wave **combine**. Then **each wave** goes on its merry way. You can **see** this if **two pulses** are sent **simultaneously** from each end of a rope.

2) The **principle of superposition** says that when two or more **waves cross**, the **resultant** displacement equals the **vector sum** of the **individual** displacements.

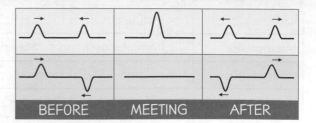

BEFORE   MEETING   AFTER

"Superposition" means "one thing on top of another thing". You can use the same idea in reverse — a complex wave can be separated out mathematically into several simple sine waves of various sizes.

Interference *can be* Constructive *or* Destructive

1) A **crest** plus a **crest** gives a **big crest**. A **trough** plus a **trough** gives a **big trough**. These are both examples of **constructive interference**.

2) A **crest** plus a **trough** of **equal size** gives... **nothing**. The two displacements **cancel each other out** completely. This is called **destructive interference**.

3) If the **crest** and the **trough** aren't the **same size**, then the destructive interference **isn't total**. For the interference to be **noticeable**, the two **amplitudes** should be **nearly equal**.

Graphically, you can superimpose waves by adding the individual displacements at each point along the x-axis, and then plotting them.

In Phase *Means In* Step *— Two Points* In Phase *Interfere* Constructively

1) Two points on a wave are **in phase** if they are both at the **same point** in the **wave cycle**. Points in phase have the **same displacement** and **velocity**.

On the graph, points **A** and **B** are **in phase**; points **A** and **C** are **out of phase**.

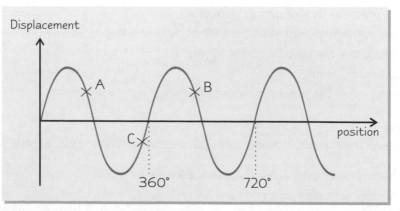

2) It's mathematically **handy** to show one **complete cycle** of a wave as an **angle of 360° (2π radians)**. **Two points** with a **phase difference** of **zero** or a **multiple of 360°** are **in phase**. **Points** with a **phase difference** of **odd-number multiples** of **180° (π radians)** are **exactly out of phase**.

3) You can also talk about two **different waves** being **in phase**. **In practice** this happens because **both** waves came from the **same oscillator**. In **other** situations there will nearly always be a **phase difference** between two waves.

Superposition and Coherence

To Get **Interference Patterns** the **Two Sources** Must Be **Coherent**

Interference **still happens** when you're observing waves of **different wavelength** and **frequency** — but it happens in a **jumble**. In order to get clear **interference patterns**, the two or more sources must be **coherent**.

In exam questions at AS, the 'fixed phase difference' is almost certainly going to be zero. The two sources will be in phase.

Two sources are **coherent** if they have the **same wavelength** and **frequency** and a **fixed phase difference** between them.

Constructive or **Destructive** Interference Depends on the **Path Difference**

1) Whether you get **constructive** or **destructive** interference at a **point** depends on how **much further one wave** has travelled than the **other wave** to get to that point.

2) The **amount** by which the path travelled by one wave is **longer** than the path travelled by the other wave is called the **path difference**.

3) At **any point an equal distance** from both sources you will get **constructive interference**. You also get constructive interference at any point where the **path difference** is a **whole number of wavelengths**. At these points the two waves are **in phase** and **reinforce** each other. But at points where the path difference is **half a wavelength**, **one and a half** wavelengths, **two and a half** wavelengths etc., the waves arrive **out of phase** and you get **destructive interference**.

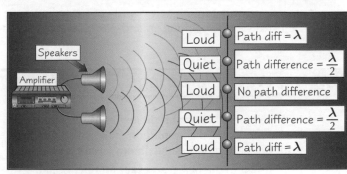

Constructive interference occurs when:

$$\text{path difference} = n\lambda \quad (\text{where } n \text{ is an integer})$$

Destructive interference occurs when:

$$\text{path difference} = \frac{(2n + 1)\lambda}{2} = (n + \tfrac{1}{2})\lambda$$

Practice Questions

Q1 Why does the principle of superposition deal with the **vector** sum of two displacements?

Q2 What happens when a crest meets a slightly smaller trough?

Q3 If two points on a wave have a phase difference of 1440°, are they in phase?

Exam Questions

Q1 (a) Two sources are coherent.
What can you say about their frequencies, wavelengths and phase difference? [2 marks]

(b) Suggest why you might have difficulty in observing interference patterns in an area affected by two waves from two sources even though the two sources are coherent. [1 mark]

Q2 Two points on an undamped wave are exactly out of phase.

(a) What is the phase difference between them, expressed in degrees? [1 mark]

(b) Compare the displacements and velocities of the two points. [2 marks]

Learn this and you'll be in a super position to pass your exam... ...I'll get my coat.

There are a few really crucial concepts here: a) interference can be constructive or destructive, b) constructive interference happens when the path difference is a whole number of wavelengths, c) the sources must be coherent.

Standing (Stationary) Waves

Standing waves are waves that... er... stand still... well, not still exactly... I mean, well... they don't go anywhere... um...

You get Standing Waves When a *Progressive Wave* is *Reflected* at a *Boundary*

A standing wave is the **superposition** of **two progressive waves** with the **same wavelength**, moving in **opposite directions**.

1) Unlike progressive waves, **no energy** is transmitted by a standing wave.

2) You can demonstrate standing waves by setting up a **driving oscillator** at one end of a **stretched string** with the other end fixed. The wave generated by the oscillator is **reflected** back and forth.

3) For most frequencies the resultant **pattern** is a **jumble**. However, if the oscillator happens to produce an **exact number of waves** in the time it takes for a wave to get to the **end** and **back again**, then the **original** and **reflected** waves **reinforce** each other.

4) At these **"resonant frequencies"** you get a **standing wave** where the **pattern doesn't move** — it just sits there, bobbing up and down. Happy, at peace with the world...

A sitting wave.

Standing Waves in *Strings* Form *Oscillating "Loops"* Separated by *Nodes*

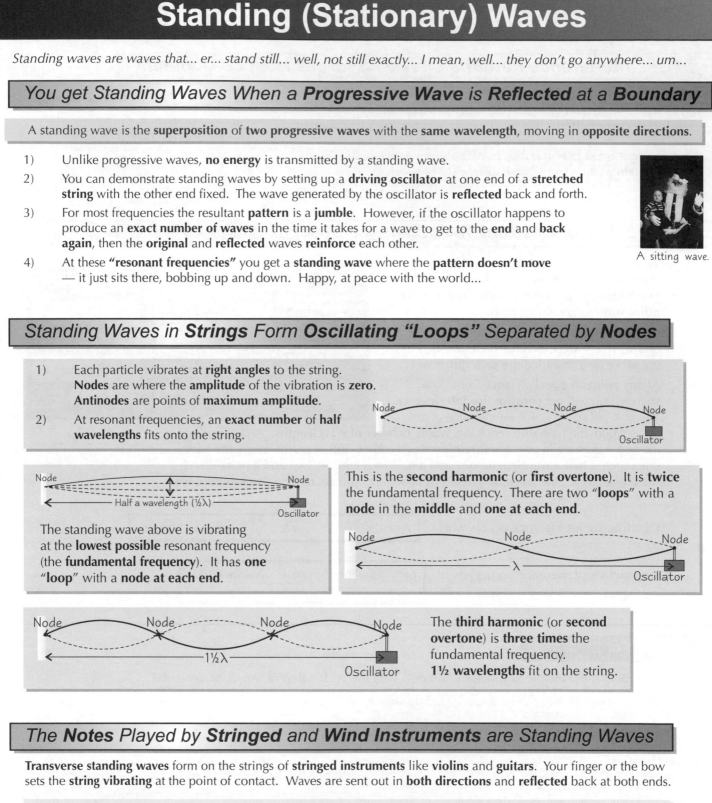

1) Each particle vibrates at **right angles** to the string.
Nodes are where the **amplitude** of the vibration is **zero**.
Antinodes are points of **maximum amplitude**.

2) At resonant frequencies, an **exact number** of **half wavelengths** fits onto the string.

The standing wave above is vibrating at the **lowest possible** resonant frequency (the **fundamental frequency**). It has **one** "loop" with a **node** at each end.

This is the **second harmonic** (or **first overtone**). It is **twice** the fundamental frequency. There are two **"loops"** with a **node** in the **middle** and **one at each end**.

The **third harmonic** (or **second overtone**) is **three times** the fundamental frequency. **1½ wavelengths** fit on the string.

The *Notes* Played by *Stringed* and *Wind Instruments* are Standing Waves

Transverse standing waves form on the strings of **stringed instruments** like **violins** and **guitars**. Your finger or the bow sets the **string vibrating** at the point of contact. Waves are sent out in **both directions** and **reflected** back at both ends.

Longitudinal Standing Waves Form in a **Wind Instrument** or Other **Air Column**

1) If a source of sound is placed at the open end of a flute, piccolo, oboe or other column of air, there will be some **frequencies** for which **resonance** occurs and a standing wave is set up.

2) If the instrument has a **closed end**, a **node** will form there. You get the lowest resonant frequency when the length, *l*, of the pipe is a **quarter wavelength**.

$$l = \frac{\lambda}{4}$$

$$l = \frac{\lambda}{2}$$

3) **Antinodes** form at the **open ends** of pipes. If both ends are open, you get the lowest resonant frequency when the length, *l*, of the pipe is a **half wavelength**.

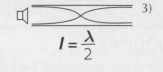

Remember, the sound waves in wind instruments are <u>longitudinal</u> — they <u>don't</u> actually look like these diagrams.

Standing (Stationary) Waves

Microwaves Reflected Off a Metal Plate Set Up a Standing Wave

Microwave standing wave apparatus ➡
You can find the **nodes** and **antinodes**
by moving the **probe** between the
transmitter and the **reflecting** plate.

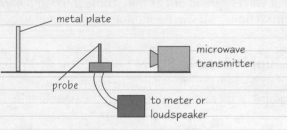

metal plate

microwave transmitter

probe

to meter or loudspeaker

You can Use Standing Waves **to Measure the** Speed of Sound

Finding the Speed of Sound in a Resonance Tube

1) You can create a closed-end pipe by placing a **hollow tube** into a measuring cylinder of water.

2) Choose a tuning fork and note down the frequency of sound it produces (it'll be stamped on the side of it).

3) Gently tap the tuning fork and hold it just above the hollow tube. The sound waves produced by the fork travel down the tube and get reflected (and form a **node**) at the air/water surface.

4) Move the tube up and down until you find the **shortest distance** between the top of the tube and the water level that the sound from the fork **resonates** at.

5) Just like with any closed pipe, this distance is a **quarter** of the wavelength of the standing sound wave.

6) The antinode of the wave actually forms slightly **above** the top of the tube — so you need to add a constant called an **end correction** to the length of your tube **before** you can work out the wavelength.

7) Once you know the **frequency** and **wavelength** of the standing sound wave, you can work out the **speed of sound** (in air), v, using the equation $v = f\lambda$.

tuning fork

$\frac{\lambda}{4}$

node

water

measuring cylinder

hollow plastic tube

Practice Questions

Q1 How do standing waves form?

Q2 At four times the fundamental frequency, how many half wavelengths fit on a violin string?

Q3 Describe an experiment to find the speed of sound in air using standing waves.

Exam Question

Q1 (a) A standing wave of three times the fundamental frequency is formed on a stretched string of length 1.2 m. Sketch a diagram showing the form of the wave. [2 marks]

(b) What is the wavelength of the standing wave? [1 mark]

(c) Explain how the amplitude varies along the string. How is that different from the amplitude of a progressive wave? [2 marks]

CGP — putting the FUN back in FUNdamental frequency...

Resonance was a big problem for the Millennium Bridge in London. The resonant frequency of the bridge was round about normal walking pace, so as soon as people started using it they set up a huge standing wave. An oversight, I feel...

Diffraction

Ripple tanks, ripple tanks — yeah.

Waves Go **Round Corners** and **Spread out** of **Gaps**

The way that **waves spread out** as they come through a **narrow gap** or go round obstacles is called **diffraction**.   **All** waves diffract, but it's not always easy to observe.

Use a **Ripple Tank** To Show Diffraction of **Water Waves**

You can make diffraction patterns in ripple tanks.
The **amount** of diffraction depends on the **wavelength** of the wave compared with the **size of the gap**.

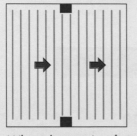   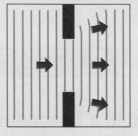   

When the gap is **a lot bigger** than the **wavelength**, diffraction is **unnoticeable**.

You get **noticeable diffraction** through a gap **several** wavelengths wide.

You get the **most** diffraction when the gap is **the same** size as the **wavelength**.

If the gap is **smaller** than the wavelength, the waves are mostly just **reflected back**.

When **sound** passes through a **doorway**, the **size of gap** and the **wavelength** are usually roughly **equal**, so **a lot** of **diffraction** occurs. That's why you have no trouble **hearing** someone through an **open door** to the next room, even if the other person is out of your **line of sight**. The reason that you can't **see** him or her is that when **light** passes through the doorway, it is passing through a **gap** around a **hundred million times bigger** than its wavelength — the amount of diffraction is **tiny**.

Demonstrate **Diffraction** in **Light** Using **Laser Light**

1) Diffraction in **light** can be demonstrated by shining a **laser light** through a very **narrow slit** onto a screen (see page 69). You can alter the amount of diffraction by changing the width of the slit.

2) You can do a similar experiment using a **white light** source instead of the laser (which is monochromatic) and a set of **colour filters**. The size of the slit can be kept constant while the **wavelength** is varied by putting different **colour filters** over the slit.

Warning. Use of coloured filters may result in excessive fun.

You Get a **Similar** Effect Around an **Obstacle**

When a wave meets an **obstacle**, you get diffraction around the edges.

Behind the obstacle is a '**shadow**', where the wave is blocked. The **wider** the obstacle compared with the wavelength of the wave, the less diffraction you get, and so the **longer** the shadow.

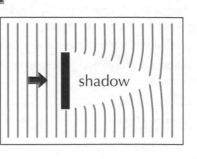

Diffraction

Diffraction is Sometimes *Useful* and Sometimes a Pain...

1) For a **loudspeaker** you want the sound to be heard as widely as possible, so you aim to **maximise** diffraction.

2) With a **microwave oven** you want to **stop** the **microwaves** diffracting out and frying your kidneys **and** you want to **let light through** so you can **see** your food. A **metal mesh** on the **door** has **gaps too small** for microwaves to diffract through, but **light** slips through because of its **tiny wavelength**.

With *Light Waves* you get a *Pattern* of *Light* and *Dark Fringes*

1) If the wavelength of a light wave is about the same size as the aperture, you get a **diffraction pattern** of light and dark fringes.

2) The pattern has a **bright central fringe** with alternating **dark and bright fringes** on either side of it.

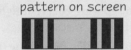

pattern on screen

light — slit — screen

You need to use a coherent light source for this experiment.

3) The **narrower** the slit, the **wider** the diffraction pattern.

You Get a *Similar Pattern* with *Electrons*

1) It's not just with **light** that you get diffraction patterns.

2) In **1927**, two American physicists, **Clinton Davisson** and **Lester Germer**, succeeded in diffracting **electrons**.

3) This was a **huge** discovery. A few years earlier, **Louis de Broglie** had **hypothesised** that electrons would show **wave-like** properties (in the same way that light can show particle-like properties — more about that in the next section), but this was the first **direct evidence** for it.

Electron diffraction patterns look like this

Practice Questions

Q1 What is diffraction?

Q2 Sketch what happens when plane waves meet an obstacle about as wide as one wavelength.

Q3 For a long time some scientists argued that light couldn't be a wave because it did not seem to diffract. Suggest why they might have got this impression.

Q4 Do all waves diffract?

Exam Question

Q1 A mountain lies directly between you and a radio transmitter.

Explain using diagrams why you can pick up long-wave radio broadcasts from the transmitter but not short-wave radio broadcasts.

[4 marks]

Even hiding behind a mountain, you can't get away from long-wave radio...

*Diffraction crops up again in particle physics, quantum physics and astronomy, so you **really** need to understand it.*

Two-Source Interference

Yeah, I know, fringe spacing doesn't really sound like a Physics topic — just trust me on this one, OK.

Demonstrating Two-Source Interference in **Water** and **Sound** is Easy

1) It's **easy** to demonstrate **two-source interference** for either **sound** or **water** because they've got **wavelengths** of a handy **size** that you can **measure**.

2) You need **coherent** sources, which means the **wavelength** and **frequency** have to be the **same**. The trick is to use the **same oscillator** to drive **both sources**. For water, one **vibrator drives two dippers**. For sound, **one oscillator** is connected to **two loudspeakers**. (See diagram on page 65.)

Demonstrating **Two-Source** Interference for **Light** is Harder

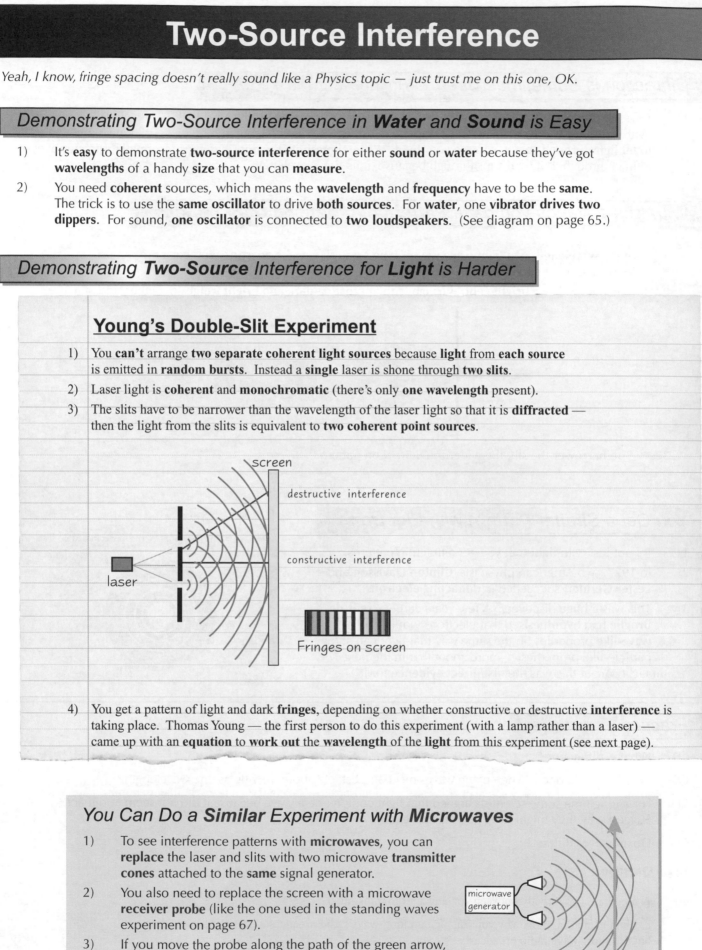

Young's Double-Slit Experiment

1) You **can't** arrange **two separate coherent light sources** because **light** from **each source** is emitted in **random bursts**. Instead a **single** laser is shone through **two slits**.

2) Laser light is **coherent** and **monochromatic** (there's only **one wavelength** present).

3) The slits have to be narrower than the wavelength of the laser light so that it is **diffracted** — then the light from the slits is equivalent to **two coherent point sources**.

screen

destructive interference

constructive interference

laser

Fringes on screen

4) You get a pattern of light and dark **fringes**, depending on whether constructive or destructive **interference** is taking place. Thomas Young — the first person to do this experiment (with a lamp rather than a laser) — came up with an **equation** to **work out** the **wavelength** of the **light** from this experiment (see next page).

You Can Do a **Similar** Experiment with **Microwaves**

1) To see interference patterns with **microwaves**, you can **replace** the laser and slits with two microwave **transmitter cones** attached to the **same** signal generator.

2) You also need to replace the screen with a microwave **receiver probe** (like the one used in the standing waves experiment on page 67).

3) If you move the probe along the path of the green arrow, you'll get an **alternating pattern** of **strong** and **weak** signals — just like the light and dark fringes on the screen.

microwave generator

probe

Two-Source Interference

Work Out the Wavelength with Young's Double-Slit Formula

1) The fringe spacing (**X**), wavelength (**λ**), spacing between slits (**d**) and the distance from slits to screen (**D**) are all related by **Young's double-slit formula**, which works for all waves (you need to know it, but not derive it).

$$\text{Fringe spacing, } X = \frac{D\lambda}{d}$$

"Fringe spacing" means the distance from the centre of one minimum to the centre of the next minimum or from the centre of one maximum to the centre of the next maximum.

2) Since the wavelength of light is so small you can see from the formula that a high ratio of **D / d** is needed to make the fringe spacing **big enough to see**.

3) Rearranging, you can use **λ = Xd / D** to **calculate the wavelength** of light.

4) The fringes are **so tiny** that it's very hard to get an **accurate value of X**. It's easier to measure across **several** fringes then **divide** by the number of **fringe widths** between them.

Always check your fringe spacing.

Young's Experiment was Evidence for the Wave Nature of Light

1) Towards the end of the **17th century**, two important **theories of light** were published — one by Isaac Newton and the other by a chap called Huygens. **Newton's** theory suggested that light was made up of tiny particles, which he called "**corpuscles**". And **Huygens** put forward a theory using **waves**.

2) The **corpuscular theory** could explain **reflection** and **refraction**, but **diffraction** and **interference** are both **uniquely** wave properties. If it could be **shown** that light showed interference patterns, that would help settle the argument once and for all.

3) **Young's** double-slit experiment (over 100 years later) provided the necessary evidence. It showed that light could both **diffract** (through the narrow slits) and **interfere** (to form the interference pattern on the screen).

Of course, this being Physics, nothing's ever simple — give it another 100 years or so and the debate would be raging again. But that can wait for the next section...

Practice Questions

Q1 In Young's experiment, why do you get a bright fringe at a point equidistant from both slits?

Q2 What does Young's experiment show about the nature of light?

Q3 Write down Young's double-slit formula.

Exam Questions

Q1 (a) The diagram on the right shows waves from two coherent light sources, S_1 and S_2.
Sketch the interference pattern, marking on constructive and destructive interference.
[2 marks]

(b) In practice if interference is to be observed, S_1 and S_2 must be slits in a screen behind which there is a source of laser light. Why? [2 marks]

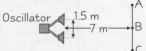

Q2 In an experiment to study sound interference, two loudspeakers are connected to an oscillator emitting sound at 1320 Hz and set up as shown in the diagram below. They are 1.5 m apart and 7 m away from the line AC. A listener moving from A to C hears minimum sound at A and C and maximum sound at B.

(a) Calculate the wavelength of the sound waves if the speed of sound in air is taken to be 330 ms⁻¹. [1 mark]

(b) Calculate the separation of points A and C. [2 marks]

Carry on Physics — this page is far too saucy...

Be careful when you're calculating the fringe width by averaging over several fringes. Don't just divide by the number of bright lines. Ten bright lines will only have nine fringe-widths between them, not ten. It's an easy mistake to make, but you have been warned... mwa ha ha ha (felt necessary, sorry).

* Sadly, Barbara Windsor was unavailable.

Diffraction Gratings

Ay... starting to get into some pretty funky stuff now. I like light experiments.

Interference Patterns Get **Sharper** When You Diffract Through **More Slits**

1) You can repeat **Young's double-slit** experiment (see p. 70) with **more than two equally spaced** slits. You get basically the **same shaped** pattern as for two slits — but the **bright bands** are **brighter** and **narrower** and the **dark areas** between are **darker**.

2) When **monochromatic light** (one wavelength) is passed through a **grating** with **hundreds** of slits per millimetre, the interference pattern is **really sharp** because there are so **many beams reinforcing** the **pattern**.

3) Sharper fringes make for more **accurate** measurements.

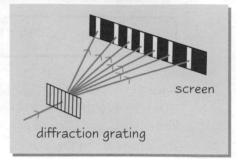
screen

diffraction grating

Monochromatic Light on a **Diffraction Grating** gives **Sharp Lines**

1) For **monochromatic** light, all the **maxima** are sharp lines. (It's different for white light — see next page.)

2) There's a line of **maximum brightness** at the centre called the **zero order** line.

3) The lines just **either side** of the central one are called **first order lines**. The **next pair out** are called **second order** lines and so on.

4) For a grating with slits a distance **d** apart, the angle between the **incident beam** and **the nth order maximum** is given by:

$$d \sin \theta = n\lambda$$

5) So by observing **d**, **θ** and **n** you can **calculate the wavelength** of the light.

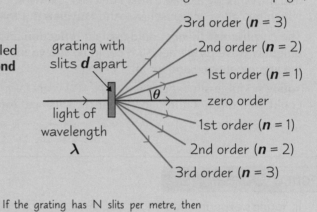

grating with slits **d** apart

light of wavelength **λ**

3rd order (**n** = 3)
2nd order (**n** = 2)
1st order (**n** = 1)
zero order
1st order (**n** = 1)
2nd order (**n** = 2)
3rd order (**n** = 3)

If the grating has N slits per metre, then the slit spacing, d, is just 1/N metres.

WHERE THE EQUATION COMES FROM:

1) At **each slit**, the incoming waves are **diffracted**. These diffracted waves then **interfere** with each other to produce an **interference pattern**.

2) Consider the **first order maximum**. This happens at the **angle** when the waves from one slit line up with waves from the **next slit** that are **exactly one wavelength** behind.

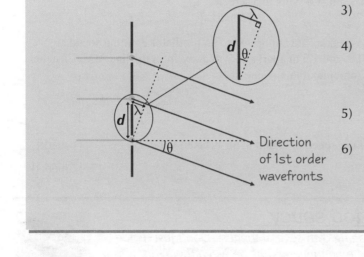

d
d
θ
Direction of 1st order wavefronts

3) Call the **angle** between the **first order maximum** and the **incoming light** θ.

4) Now, look at the **triangle** highlighted in the diagram. The angle is θ (using basic geometry), **d** is the slit spacing and the **path difference** is λ.

5) So, for the first maximum, using trig:
$$d \sin \theta = \lambda$$

6) The other maxima occur when the path difference is 2λ, 3λ, 4λ, etc. So to make the equation **general**, just replace λ with **n**λ, where **n** is an integer — the **order** of the maximum.

Diffraction Gratings

You can Draw General Conclusions from d sin θ = nλ

1) If λ is **bigger**, sin θ is **bigger**, and so θ is **bigger**. This means that the larger the **wavelength**, the more the pattern will **spread out**.

2) If d is **bigger**, sin θ is **smaller**. This means that the **coarser** the **grating**, the **less** the pattern will **spread out**.

3) Values of sin θ greater than **1** are **impossible**. So if for a certain n you get a result of **more than 1** for sin θ you know that that order **doesn't exist**.

Shining White Light Through a Diffraction Grating Produces Spectra

1) **White light** is really a **mixture** of **colours**. If you **diffract** white light through a **grating** then the patterns due to **different wavelengths** within the white light are **spread out** by **different** amounts.

2) Each **order** in the pattern becomes a **spectrum**, with **red** on the **outside** and **violet** on the **inside**. The **zero order maximum** stays **white** because all the wavelengths just pass straight through.

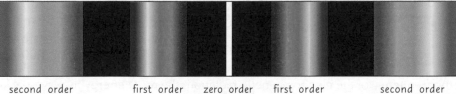

second order   first order   zero order   first order   second order
(white)

Astronomers and **chemists** often need to study spectra to help identify elements. They use diffraction gratings rather than prisms because they're **more accurate**.

Practice Questions

Q1 How is the diffraction grating pattern for white light different from the pattern for laser light?

Q2 What difference does it make to the pattern if you use a finer grating?

Q3 What equation is used to find the angle between the nth order maximum and the incident beam for a diffraction grating?

Exam Questions

Q1 Yellow laser light of wavelength 600 nm (6×10^{-7} m) is transmitted through a diffraction grating of 4×10^5 lines per metre.

(a) At what angle to the normal are the first and second order bright lines seen?   [4 marks]

(b) Is there a fifth order line?   [1 mark]

Q2 Visible, monochromatic light is transmitted through a diffraction grating of 3.7×10^5 lines per metre. The first order maximum is at an angle of 14.2° to the incident beam.

Find the wavelength of the incident light.   [2 marks]

Ooooooooooooo — pretty patterns...

Yes, it's the end of another beautiful section — woohoo. Three important points for you to take away — the more slits you have, the sharper Draw image, one lovely equation to learn and white light makes a pretty spectrum. Make sure you get everything in this section — there's some good stuff coming up in the next one and I wouldn't want you to be distracted.

Light — Wave or Particle

You probably already thought light was a bit weird — but oh no... being a wave that travels at the fastest speed possible isn't enough for light — it has to go one step further and act like a particle too...

Light Behaves Like a **Wave**... or a **Stream of Particles**

1) In the **late nineteenth century**, if you asked what light was, scientists would happily show you lots of nice experiments showing how light must be a **wave** (see Unit 2: Section 2).

2) Then came the **photoelectric effect** (p. 76), which mucked up everything. The only way you could explain this was if light acted as a **particle** — called a **photon**.

A **Photon** is a **Quantum** of **EM Radiation**

1) When Max Planck was investigating **black body radiation** (don't worry — you don't need to know about that just yet), he suggested that **EM waves** can **only** be **released** in **discrete packets**, called **quanta**. A single packet of **EM radiation** is called a **quantum**.

The **energy carried** by one of these **wave-packets** had to be:

$$E = hf = \frac{hc}{\lambda}$$

where h = Planck's constant = 6.63×10^{-34} Js,
f = frequency (Hz), λ = wavelength (m)
and c = speed of light in a vacuum = 3.00×10^8 ms^{-1}

2) So, the **higher** the **frequency** of the electromagnetic radiation, the more **energy** its wave-packets carry.

3) **Einstein** went **further** by suggesting that **EM waves** (and the energy they carry) can only **exist** in discrete packets. He called these wave-packets **photons**.

4) He believed that a photon acts as **particle**, and will either transfer **all** or **none** of its energy when interacting with another particle, like an electron.

Photon Energies are Usually Given in **Electronvolts**

1) The **energies involved** when you're talking about photons are **so tiny** that it makes sense to use a more **appropriate unit** than the **joule**. Bring on the **electronvolt** ...

2) When you **accelerate** an electron between two electrodes, it transfers some electrical potential energy (eV) into kinetic energy.

$$eV = \frac{1}{2}mv^2$$

e is the charge on an electron: 1.6×10^{-19} C.

3) An electronvolt is defined as:

> The **kinetic energy gained** by an **electron** when it is **accelerated** through a **potential difference** of **1 volt**.

4) So 1 electron volt = $e \times V$ = 1.6×10^{-19} C $\times$ 1 JC^{-1}. $\implies$ 1 eV = 1.6×10^{-19} J

Threshold Voltage is Used to Find **Planck's Constant**

1) Planck's constant comes up everywhere — but it's not just some random number plucked out of the air. You can find its value by doing a simple experiment with **light-emitting diodes** (**LED**s).

2) Current will only pass through an LED after a **minimum voltage** is placed across it — the **threshold voltage** V_0.

3) This is the voltage needed to give the electrons the **same energy** as a photon emitted by the LED. **All** of the electron's **kinetic energy** after it is accelerated over this potential difference is **transferred** into a **photon**.

$$E = \frac{hc}{\lambda} = eV_0 \Rightarrow h = \frac{(eV_0)\lambda}{c}$$

4) So by finding the threshold voltage for a particular wavelength LED, you can calculate Planck's constant.

Light — Wave or Particle

You can Use LEDs to Calculate Planck's Constant

You've just seen the **theory** of how to find **Planck's constant** — now it's time for the **practicalities**.

Experiment to Measure Planck's Constant

6V

1) Connect an LED of known wavelength in the electrical circuit shown.

2) Start off with no current flowing through the circuit, then adjust the variable resistor until a current just begins to flow through the circuit.

3) Record the voltage (V_0) across the LED, and the wavelength of light the LED emits.

4) Repeat this experiment with a number of LEDs that emit different optical wavelengths.

mA

Voltage (V)

5) Plot a graph of threshold voltages (V_0) against $1/\lambda$ (where λ is the wavelength of light emitted by the LED in metres).

6) You should get a straight line graph with a gradient of hc/e — which you can then use to find the value of h.

E.g.

$$\text{gradient} = \frac{hc}{e} = 1.24 \times 10^{-6},$$

$$\text{so } h = \frac{1.24 \times 10^{-6} e}{c} = \frac{(1.24 \times 10^{-6}) \times (1.6 \times 10^{-19})}{3 \times 10^8}$$

$$= 6.6 \times 10^{-34} \text{ Js (2 s.f.)}$$

0      $1/\lambda$ (1/m)

Practice Questions

Q1 Give two different ways to describe the nature of light.

Q2 Write down the two formulas you can use to find the energy of a photon. Include the meanings of all the symbols you use.

Q3 What is an electronvolt? What is 1 eV in joules?

Exam Question

$c = 3.00 \times 10^8 \, ms^{-1}$

Q1 An LED is tested and found to have a threshold voltage of 1.70 V.

(a) Find the energy of the photons emitted by the LED. Give your answer in joules. [2 marks]
(b) The LED emits light with a wavelength of 700 nm.
Use your answer from a) to calculate the value of Planck's constant. [2 marks]

Millions of light particles are hitting your retinas as you read this... PANIC...

I hate it in physics when they tell you lies, make you learn it, and just when you've got to grips with it they tell you it was all a load of codswallop. This is the real deal folks — light isn't just the nice wave you've always known...

The Photoelectric Effect

The photoelectric effect was one of the original troublemakers in the light-is-it-a-wave-or-a-particle problem...

Shining Light on a Metal can Release Electrons

If you shine **light** of a **high enough frequency** onto the **surface of a metal**,
it will **emit electrons**. For **most** metals, this **frequency** falls in the **U.V.** range.

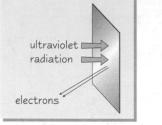

ultraviolet
radiation

electrons

1) **Free electrons** on the **surface** of the metal **absorb energy** from the light, making them **vibrate**.

2) If an electron **absorbs enough** energy, the **bonds** holding it to the metal **break** and the electron is **released**.

3) This is called the **photoelectric effect** and the electrons emitted are called **photoelectrons**.

You don't need to know the details of any experiments on this — you just need to learn the three main conclusions:

Conclusion 1	For a given metal, **no photoelectrons are emitted** if the radiation has a frequency **below** a certain value — called the **threshold frequency**.
Conclusion 2	The photoelectrons are emitted with a variety of kinetic energies ranging from zero to some maximum value. This value of **maximum kinetic energy** increases with the **frequency** of the radiation, and is **unaffected** by the **intensity** of the radiation.
Conclusion 3	The **number** of photoelectrons emitted per second is **proportional** to the **intensity** of the radiation.

These are the two that had scientists puzzled. They can't be explained using wave theory.

The Photoelectric Effect Couldn't be Explained by Wave Theory

According to wave theory:

1) For a particular frequency of light, the **energy** carried is **proportional** to the **intensity** of the beam.

2) The energy carried by the light would be **spread evenly** over the wavefront.

3) **Each** free electron on the surface of the metal would gain a **bit of energy** from each incoming wave.

4) Gradually, each electron would gain **enough energy** to leave the metal.

SO... If the light had a **lower frequency** (i.e. was carrying less energy) it would take **longer** for the electrons to gain enough energy — but it would happen eventually. There is **no explanation** for the **threshold frequency**.

The **higher the intensity** of the wave, the **more energy** it should transfer to each electron — the kinetic energy should increase with **intensity**. There's **no explanation** for the **kinetic energy** depending only on the **frequency**.

The Photon Model Explained the Photoelectric Effect Nicely

According to the photon model (see page 74)**:**

1) When light hits its surface, the metal is **bombarded** by photons.

2) If one of these photons **collides** with a free electron, the electron will gain energy equal to *hf*.

Before an electron can **leave** the surface of the metal, it needs enough energy to **break the bonds holding it there**. This energy is called the **work function energy** (symbol ϕ, phi) and its **value** depends on the **metal**.

The Photoelectric Effect

It Explains the Threshold Frequency...

1) If the energy **gained** from the photon is **greater** than the **work function energy**, the electron can be **emitted**.

2) If it **isn't**, the electron will just **shake about a bit**, then release the energy as another photon. The metal will heat up, but **no electrons** will be emitted.

3) Since for **electrons** to be released, $hf \geq \phi$, the **threshold frequency** must be:

$$f = \frac{\phi}{h}$$

In theory, if a second photon hit an electron before it released the energy from the first, it could gain enough to leave the metal. This would have to happen very quickly though. An electron releases any excess energy after about 10^{-8} s. That's 0.000 000 01 s — safe to say, the chances of that happening are pretty slim.

... and the Maximum Kinetic Energy

1) The **energy transferred** to an electron is **hf**.

2) The **kinetic energy** it will be carrying when it **leaves** the metal will be h*f* **minus** any energy it's **lost** on the way out (there are loads of ways it can do that, which explains the **range** of energies).

3) The **minimum** amount of energy it can lose is the **work function energy**, so the **maximum kinetic energy** is given by the equation:

$$hf = \phi + \frac{1}{2}mv_{max}^{2}$$

4) The **kinetic energy** of the electrons is **independent of the intensity**, because they can **only absorb one photon** at a time.

Practice Questions

Q1 Describe an experiment that demonstrates the photoelectric effect.

Q2 What is meant by the threshold frequency?

Q3 Write down the equation that relates the work function of a metal and the threshold frequency.

Q4 Write an equation that relates the maximum kinetic energy of a photoelectron released from a metal surface and the frequency of the incident light on the surface.

Exam Questions

$h = 6.63 \times 10^{-34}\ Js$

Q1 The work function of calcium is 2.9 eV.
Find the threshold frequency of radiation needed for the photoelectric effect to take place. [2 marks]

Q2 The surface of a copper plate is illuminated with monochromatic ultraviolet light, with a frequency of 2.0×10^{15} Hz. The work function for copper is 4.7 eV.
(a) Find the energy in eV carried by one ultraviolet photon. [3 marks]
(b) Find the maximum kinetic energy of a photoelectron emitted from the copper surface. [2 marks]

Q3 Explain why the photoelectric effect only occurs after the incident light has reached a certain frequency. [2 marks]

I'm so glad we got that all cleared up...

Well, that's about as hard as it gets at AS. The most important bits here are why wave theory doesn't explain the phenomenon, and why the photon theory does. A good way to learn conceptual stuff like this is to try to explain it to someone else. You'll get the formulas in your handy data book, but it's probably a good idea to learn them too...

Energy Levels and Photon Emission

Electrons in Atoms Exist in Discrete Energy Levels

1) **Electrons** in an **atom** can **only exist** in certain **well-defined energy levels**. Each level is given a **number**, with **n = 1** representing the **ground state**.

2) Electrons can **move down** an energy level by **emitting** a **photon**.

3) Since these **transitions** are between **definite energy levels**, the **energy** of **each photon** emitted can **only** take a **certain allowed value**.

4) The diagram on the right shows the **energy levels** for **atomic hydrogen**.

5) On the diagram, energies are labelled in both **joules** and **electonvolts** for **comparison's** sake.

6) The **energy** carried by each **photon** is **equal** to the **difference in energies** between the **two levels**. The equation below shows a **transition** between levels **n = 2** and **n = 1**:

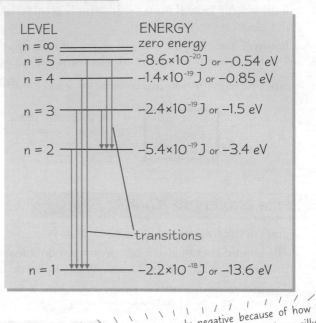

LEVEL	ENERGY
n = ∞	zero energy
n = 5	-8.6×10^{-20} J or -0.54 eV
n = 4	-1.4×10^{-19} J or -0.85 eV
n = 3	-2.4×10^{-19} J or -1.5 eV
n = 2	-5.4×10^{-19} J or -3.4 eV
n = 1	-2.2×10^{-18} J or -13.6 eV

transitions

$$\Delta E = E_2 - E_1 = hf = \frac{hc}{\lambda}$$

The energies are only negative because of how "zero energy" is defined. Just one of those silly convention things — don't worry about it.

Hot Gases Produce Line Emission Spectra

1) If you heat a gas to a high temperature, many of it's electrons move to higher energy levels.

2) As they fall back to the ground state, these electrons emit energy as photons.

3) If you **split** the light from a **hot gas** with a **prism** or a **diffraction grating** (see pages 72-73), you get a **line spectrum**. A line spectrum is seen as a **series** of **bright lines** against a **black background**, as shown below.

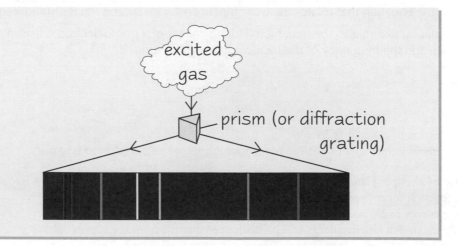

excited gas

prism (or diffraction grating)

4) Each **line** on the spectrum corresponds to a **particular wavelength** of light **emitted** by the source. Since only **certain photon energies** are **allowed**, you only see the **corresponding wavelengths**.

Energy Levels and Photon Emission

Shining *White Light* through a *Cool Gas* gives an *Absorption Spectrum*

Continuous Spectra Contain All Possible Wavelengths

1) The **spectrum** of **white light** is **continuous**.
2) If you **split** the **light** up with a **prism**, the **colours** all **merge** into each other — there **aren't** any **gaps** in the spectrum.
3) **Hot things** emit a **continuous spectrum** in the visible and infrared.

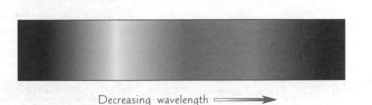

Decreasing wavelength ⟹

Cool Gases Remove Certain Wavelengths from the Continuous Spectrum

1) You get a **line absorption spectrum** when **light** with a **continuous spectrum** of **energy** (white light) passes through a cool gas.
2) At **low temperatures**, **most** of the **electrons** in the **gas atoms** will be in their **ground states**.
3) **Photons** of the **correct wavelength** are **absorbed** by the **electrons** to **excite** them to **higher energy levels**.
4) These **wavelengths** are then **missing** from the **continuous spectrum** when it **comes out** the other side of the gas.
5) You see a **continuous spectrum** with **black lines** in it corresponding to the **absorbed wavelengths**.
6) If you **compare** the **absorption** and **emission** spectra of a **particular gas**, the **black lines** in the **absorption spectrum match up** to the **bright lines** in the **emission spectrum**.

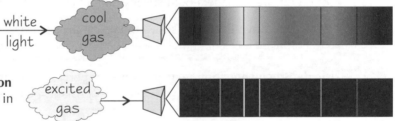

Practice Questions

Q1   Describe line absorption and line emission spectra. How are these two types of spectra produced?

Q2   Use the size of the energy level transitions involved to explain how the coating on a fluorescent tube converts UV into visible light.

Exam Question

$e = 1.6 \times 10^{-19}$ C

Q1   An electron is accelerated through a potential difference of 12.1 V.

(a)   How much kinetic energy has it gained in (i) eV and (ii) joules?   [2 marks]

(b)   This electron hits a hydrogen atom and excites it.
 (i)   Explain what is meant by excitation.   [1 mark]
 (ii)   Using the energy values on the right, work out to which energy level the electron from the hydrogen atom is excited.   [1 mark]
 (iii)   Calculate the energies of the three photons that might be emitted as the electron returns to its ground state.   [3 marks]

```
n = 5 ——————— − 0.54 eV
n = 4 ——————— − 0.85 eV
n = 3 ——————— − 1.5 eV
n = 2 ——————— − 3.4 eV
n = 1 ——————— − 13.6 eV
```

I can honestly say I've never got so excited that I've produced light...

This is heavy stuff, it really is. Quite interesting though, as I was just saying to Dom a moment ago. He's doing a psychology book. Psychology's probably quite interesting too — and easier. But it won't help you become an astrophysicist.

Wave-Particle Duality

Is it a wave? Is it a particle? No, it's a wave. No, it's a particle. No it's not, it's a wave. No don't be daft, it's a particle. (etc.)

Interference and Diffraction show Light as a Wave

1) Light produces **interference** and **diffraction** patterns — **alternating bands** of **dark** and **light**.

2) These can **only** be explained using **waves interfering constructively** (when two waves overlap in phase) or **interfering destructively** (when the two waves are out of phase). (See p.65.)

The Photoelectric Effect Shows Light Behaving as a Particle

1) **Einstein** explained the results of **photoelectricity** experiments (see p.76) by thinking of the **beam of light** as a series of **particle-like photons**.

2) If a **photon** of light is a **discrete** bundle of energy, then it can **interact** with an **electron** in a **one-to-one way**.

3) **All** the **energy** in the **photon** is **given** to one **electron**.

De Broglie Came up With the Wave-Particle Duality Theory

1) Louis de Broglie made a **bold suggestion** in his **PhD thesis**:

> If 'wave-like' light showed **particle properties** (photons), 'particles' like **electrons** should be expected to show **wave-like properties**.

2) The **de Broglie equation** relates a **wave property** (wavelength, λ) to a **moving particle property** (**momentum**, mv). h = Planck's constant = 6.63×10^{-34} Js.

$$\lambda = \frac{h}{mv}$$

I'm not impressed — this is just speculation. What do you think Dad?

3) The **de Broglie wave** of a particle can be interpreted as a '**probability wave**'. (The probability of finding a particle at a point is directly proportional to the square of the amplitude of the wave at that point — but you don't need to know that for your exam.)

4) Many physicists at the time **weren't very impressed** — his ideas were just **speculation**. But later experiments **confirmed** the wave nature of electrons.

Electron Diffraction shows the Wave Nature of Electrons

1) **Diffraction patterns** are observed when **accelerated electrons** in a vacuum tube **interact** with the **spaces** in a graphite **crystal**.

2) This **confirms** that electrons show **wave-like** properties.

3) According to wave theory, the **spread** of the **lines** in the diffraction pattern **increases** if the **wavelength** of the wave is **greater**.

4) In electron diffraction experiments, a **smaller accelerating voltage**, i.e. **slower** electrons, gives **widely spaced** rings.

5) **Increase** the **electron speed** and the diffraction pattern circles **squash together** towards the **middle**. This fits in with the **de Broglie** equation above — if the **velocity** is **higher**, the wavelength is **shorter** and the spread of lines is **smaller**.

> In general, λ for **electrons** accelerated in a **vacuum tube** is about the same size as **electromagnetic waves** in the **X-ray** part of the spectrum.

Wave-Particle Duality

Particles Don't show Wave-Like Properties All the Time

You **only** get **diffraction** if a particle interacts with an object of about the **same size** as its **de Broglie wavelength**.

A **tennis ball**, for example, with **mass 0.058 kg** and **speed 100 ms⁻¹** has a **de Broglie wavelength** of **10⁻³⁴ m**. That's **10¹⁹ times smaller** than the **nucleus** of an **atom**! There's nothing that small for it to interact with.

Example An electron of mass 9.11×10^{-31} kg is fired from an electron gun at 7×10^6 ms⁻¹. What size object will the electron need to interact with in order to diffract?

Momentum of electron = mv = 6.38×10^{-24} kg ms⁻¹

$\lambda = h/mv = 6.63 \times 10^{-34} / 6.38 \times 10^{-24} = $ 1×10^{-10} m

Only crystals with atom layer spacing around this size are likely to cause the diffraction of this electron.

A **shorter wavelength** gives **less diffraction effects**. This fact is used in the **electron microscope**.

Diffraction effects **blur detail** on an image. If you want to **resolve tiny detail** in an **image**, you need a **shorter wavelength**. **Light** blurs out detail more than 'electron-waves' do, so an **electron microscope** can resolve **finer detail** than a **light microscope**. They can let you look at things as tiny as a single strand of DNA... which is nice.

Practice Questions

Q1 Which observations show light to have a 'wave-like' character?

Q2 Which observations show light to have a 'particle' character?

Q3 What happens to the de Broglie wavelength of a particle if its velocity increases?

Q4 Which observations show electrons to have a 'wave-like' character?

Exam Questions

$h = 6.63 \times 10^{-34}$ Js ; $c = 3.00 \times 10^8$ ms⁻¹ ; electron mass = 9.11×10^{-31} kg ; proton mass = $1840 \times$ electron mass

Q1 (a) State what is meant by the wave-particle duality of electromagnetic radiation. [1 mark]

   (b) (i) Calculate the energy in joules and in electronvolts of a photon of wavelength 590 nm. [3 marks]

      (ii) Calculate the speed of an electron which will have the same wavelength as the photon in (b)(i). [2 marks]

Q2 Electrons travelling at a speed of 3.5×10^6 ms⁻¹ exhibit wave properties.

   (a) Calculate the wavelength of these electrons. [2 marks]

   (b) Calculate the speed of protons which would have the same wavelength as these electrons. [2 marks]

   (c) Both electrons and protons were accelerated from rest by the same potential difference. Explain why they will have different wavelengths. (Hint: if they're accelerated by the same p.d., they have the same K.E.) [3 marks]

Q3 An electron is accelerated through a potential difference of 6.0 kV.

   (a) Calculate its kinetic energy in joules, assuming no energy is lost in the process. [2 marks]

   (b) Using the data above, calculate the speed of the electron. [2 marks]

   (c) Calculate the de Broglie wavelength of the electron. [2 marks]

Don't hide your wave-particles under a bushel...

*Right — I think we'll all agree that quantum physics is a wee bit strange when you come to think about it. What it's saying is that electrons and photons aren't really waves, and they aren't really particles — they're **both**... at the same time. It's what quantum physicists like to call a 'juxtaposition of states'. Well they would, wouldn't they...*

Error Analysis

Science is all about getting good evidence to test your theories... and part of that is knowing how good the results from an experiment are. Physicists always have to include the uncertainty in a result, so you can see the range the actual value probably lies within. Dealing with error and uncertainty is an important skill, so those pesky examiners like to sneak in a couple of questions about it... but if you know your stuff you can get some easy marks.

Nothing *is* Certain

1) **Every** measurement you take has an **experimental uncertainty**. Say you've done something outrageous like measure the length of a piece of wire with a centimetre ruler. You might think you've measured its length as 30 cm, but at **best** you've probably measured it to be 30 **± 0.5** cm. And that's without taking into account any other errors that might be in your measurement...

2) The ± bit gives you the **range** in which the **true** length (the one you'd really like to know) probably lies — 30 ± 0.5 cm tells you the true length is very likely to lie in the range of 29.5 to 30.5 cm.

3) The smaller the uncertainty, the nearer your value must be to the true value, so the more **accurate** your result.

4) There are **two types** of **error** that cause experimental uncertainty:

Random errors

1) No matter how hard you try, you **can't get rid** of random errors.

2) They can just be down to **noise**, or be that you're measuring a **random process** such as nuclear radiation emission.

3) You get random error in **any** measurement.
If you measured the length of a wire 20 times, the chances are you'd get a **slightly different** value each time, e.g. due to your head being in a slightly different position when reading the scale.

4) It could be that you just can't keep controlled variables **exactly** the same throughout the experiment.

5) Or it could just be the wind was blowing in the wrong direction at the time...

Systematic errors

1) You get systematic errors not because you've made a mistake in a measurement — but because of the **apparatus** you're using, or your experimental method, e.g. using an inaccurate clock.

2) The problem is often that you **don't know they're there**. You've got to spot them first to have any chance of correcting for them.

3) Systematic errors usually **shift** all of your results to be too high or too low by the **same amount**. They're annoying, but there are things you can do to reduce them if you manage to spot them...

You Need to Know How to *Improve Measurements*

Lorraine thought getting an uncertainty of ± 0.1 A deserved a victory dance.

There are a few different ways you can **reduce** the uncertainty in your results:

Repeating measurements — by repeating a measurement **several times** and **averaging**, you reduce the **random uncertainty** in your result. The **more** measurements you average over, the **less error** you're likely to have.

Use higher precision apparatus — the **more precisely** you can measure something, the **less random error** there is in the measurement. So if you use more precise equipment — e.g. swapping a millimetre ruler for a micrometer to measure the diameter of a wire — you can instantly cut down the **random error** in your experiment.

Calibration — you can calibrate your apparatus by measuring a **known value**. If there's a **difference** between the **measured** and **known** value, you can use this to **correct** the inaccuracy of the apparatus, and so reduce your **systematic error**.

Error Analysis

You can **Calculate** the **Percentage Uncertainty** in a **Measurement**

1) You might get asked to work out the percentage uncertainty in a measurement.

2) It's just working out a percentage, so nothing too tricky. It's just that sometimes you can get **the fear** as soon as you see the word uncertainty... but just keep your cool and you can pick up some easy marks.

> **Example**
>
> Tom finds the resistance of a filament lamp to be **5.0 ± 0.4 Ω**.
> Find the percentage uncertainty in Tom's measurement.
>
> > Just divide the **absolute uncertainty** by the value for the resistance:
> >
> > $$\text{Percentage uncertainty} = \frac{0.4}{5.0} \times 100 = \mathbf{8\%}$$

You can **Estimate Values** by **Averaging**

You might be given a graph of information showing the results for many **repetitions** of the **same** experiment, and asked to estimate the true value and give an uncertainty in that value. Yuk.

Here's how to go about it:

1) Estimate the true value by **averaging** the results you've been given.
(Make sure you state whatever average it is you take, otherwise you might not get the mark.)

2) To get the uncertainty, you just need to look **how far away** from your average value the maximum and minimum values in the graph you've been given are.

> **Example — Estimating the resistance of a component**
>
> A class measure the resistance of a component and record their results on the bar chart shown on the right.
>
> Estimate the resistance of the component, giving a suitable range of uncertainty in your answer.
>
>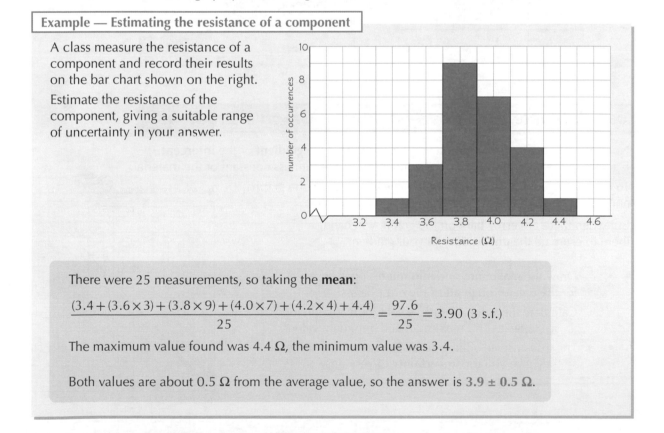
>
> > There were 25 measurements, so taking the **mean**:
> >
> > $$\frac{(3.4 + (3.6 \times 3) + (3.8 \times 9) + (4.0 \times 7) + (4.2 \times 4) + 4.4)}{25} = \frac{97.6}{25} = 3.90 \text{ (3 s.f.)}$$
> >
> > The maximum value found was 4.4 Ω, the minimum value was 3.4.
> >
> > Both values are about 0.5 Ω from the average value, so the answer is **3.9 ± 0.5 Ω**.

Random error in your favour — collect £200...

These pages should give you a fair idea of how to deal with errors... which are rather annoyingly in everything. Even if you're lucky enough to not get tested on this sort of thing in the exam, it's really useful to know for your lab coursework. Have a look at the last page for how to show errors on graphs...

Errors on Graphs

No, you can't use errors and uncertainty as an excuse for getting your graph wrong.

Error Bars to Show Uncertainty on a Graph

1) Most of the time in science, you work out the uncertainty in your **final result** using the uncertainty in **each measurement** you make.

2) When you're plotting a graph, you show the uncertainty in a value by using **error bars** to show the range the point is likely to lie in.

3) You probably won't get asked to **plot** any error bars (phew...) — but you might need to **read off** a graph that has them.

Example

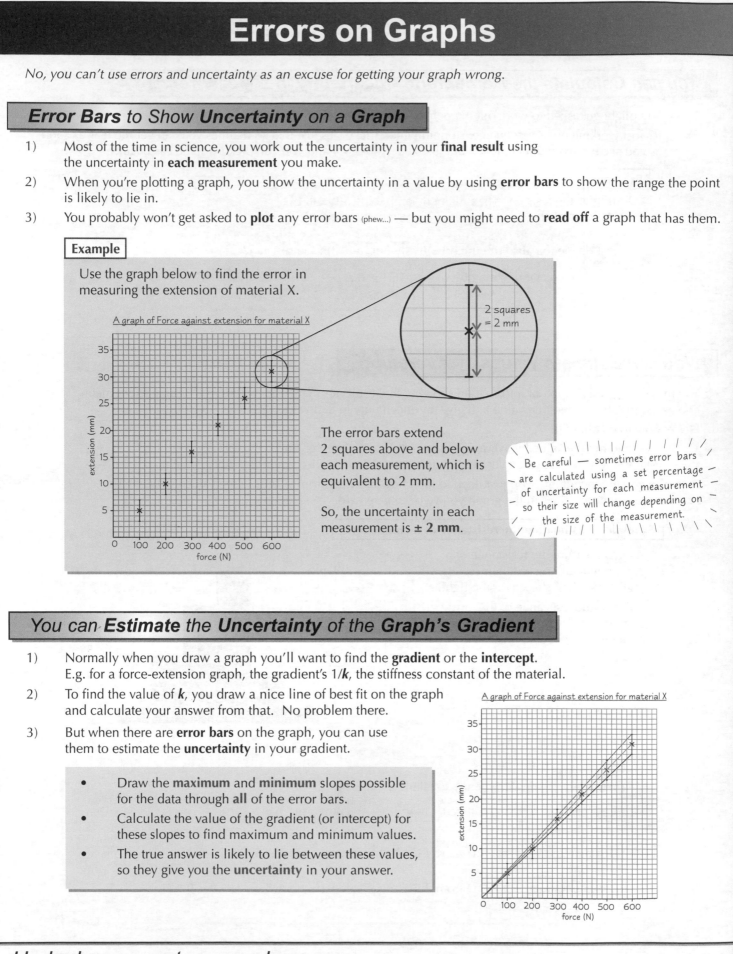

Use the graph below to find the error in measuring the extension of material X.

A graph of Force against extension for material X

2 squares = 2 mm

The error bars extend 2 squares above and below each measurement, which is equivalent to 2 mm.

So, the uncertainty in each measurement is ± **2 mm**.

Be careful — sometimes error bars are calculated using a set percentage of uncertainty for each measurement so their size will change depending on the size of the measurement.

You can Estimate the Uncertainty of the Graph's Gradient

1) Normally when you draw a graph you'll want to find the **gradient** or the **intercept**. E.g. for a force-extension graph, the gradient's $1/k$, the stiffness constant of the material.

2) To find the value of k, you draw a nice line of best fit on the graph and calculate your answer from that. No problem there.

3) But when there are **error bars** on the graph, you can use them to estimate the **uncertainty** in your gradient.

- Draw the **maximum** and **minimum** slopes possible for the data through **all** of the error bars.

- Calculate the value of the gradient (or intercept) for these slopes to find maximum and minimum values.

- The true answer is likely to lie between these values, so they give you the **uncertainty** in your answer.

A graph of Force against extension for material X

Hedgehog nougat — error bar...

Errors and uncertainty aren't a Physicist's cop-out — they're a really useful way of saying how confident you are that you've got the right answer. Getting slightly different results from your mate doesn't mean you've done something wrong, it probably just means there's some factor you can't control that's causing random errors in your results.

A2-Level
Physics

Exam Board: OCR A

Momentum and Impulse

We're going to kick off with a couple of pages about linear momentum — that's momentum in a straight line (not a circle or anything complicated like that). You'll have met momentum before so it shouldn't be too hard.

Understanding **Momentum** helps you do **Calculations** on **Collisions**

The **momentum** of an object depends on two things — its **mass** and **velocity**.
The **product** of these two values is the momentum of the object.

$$\text{momentum} = \text{mass} \times \text{velocity}$$

or in symbols: $p \text{ (in kg ms}^{-1}) = m \text{ (in kg)} \times v \text{ (in ms}^{-1})$

Hands up if you want to
learn some physics.

Remember, momentum is a **vector quantity**, so just like velocity, it has size and direction.

Momentum is always **Conserved**

1) Assuming **no external forces** act, momentum is always **conserved**.

2) This means the **total momentum** of two objects **before** they collide **equals** the total momentum **after** the collision.

3) This is really handy for working out the **velocity** of objects after a collision (as you do...):

Example   A skater of mass 75 kg and velocity 4 ms⁻¹ collides with a stationary skater of mass 50 kg.
The two skaters join together and move off in the same direction. Calculate their velocity after impact.

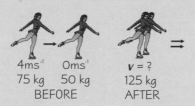

4ms⁻¹   0ms⁻¹   v = ?
75 kg   50 kg   125 kg
BEFORE   AFTER

Before you start a momentum calculation,
always draw a quick sketch.

Momentum of skaters before = Momentum of skaters after
$$(75 \times 4) + (50 \times 0) = 125v$$
$$300 = 125v$$
$$\text{So } v = 2.4 \text{ ms}^{-1}$$

4) The same principle can be applied in **explosions**. E.g. if you fire an **air rifle**, the **forward momentum** gained by the pellet **equals** the **backward momentum** of the rifle, and you feel the rifle recoiling into your shoulder.

Example   A bullet of mass 0.005 kg is shot from a rifle at a speed of 200 ms⁻¹.
The rifle has a mass of 4 kg. Calculate the velocity at which the rifle recoils.

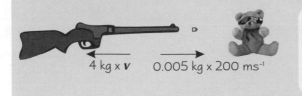

4 kg × v   0.005 kg × 200 ms⁻¹

Momentum before explosion = Momentum after explosion
$$0 = (0.005 \times 200) + (4 \times v)$$
$$0 = 1 + 4v$$
$$v = -0.25 \text{ ms}^{-1}$$

5) In reality, collisions and explosions usually happen in more than one dimension.
Don't worry though — you'll only be given problems to solve in **one dimension**.

Momentum and Impulse

Collisions can be Elastic or Inelastic

An **elastic collision** is one where **momentum** is **conserved** and **kinetic energy** is **conserved** — i.e. no energy is dissipated as heat, sound, etc. If a collision is **inelastic** it means that some of the kinetic energy is converted into other forms during the collision. But **momentum is always conserved.**

Example   A toy lorry (mass 2 kg) travelling at 3 ms⁻¹ crashes into a smaller toy car (mass 800 g), travelling in the same direction at 2 ms⁻¹. The velocity of the lorry after the collision is 2.6 ms⁻¹ in the same direction. Calculate the new velocity of the car and the total kinetic energy before and after the collision.

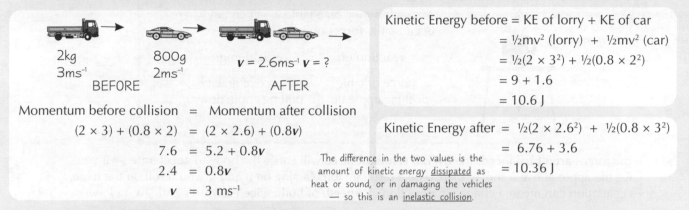

2kg
3ms⁻¹
800g
2ms⁻¹
BEFORE

$v = 2.6ms^{-1}$ $v = ?$
AFTER

Momentum before collision = Momentum after collision
$(2 \times 3) + (0.8 \times 2)$ = $(2 \times 2.6) + (0.8v)$
7.6 = $5.2 + 0.8v$
2.4 = $0.8v$
v = 3 ms^{-1}

The difference in the two values is the amount of kinetic energy <u>dissipated</u> as heat or sound, or in damaging the vehicles — so this is an <u>inelastic collision</u>.

Kinetic Energy before = KE of lorry + KE of car
= ½mv² (lorry) + ½mv² (car)
= ½(2 × 3²) + ½(0.8 × 2²)
= 9 + 1.6
= 10.6 J

Kinetic Energy after = ½(2 × 2.6²) + ½(0.8 × 3²)
= 6.76 + 3.6
= 10.36 J

Impulse = Change in Momentum

1) Newton's second law says **force = rate of change of momentum** (see page 88), or $F = (mv - mu) \div t$

2) **Rearranging** Newton's 2nd law gives: ⟹
   Impulse is defined as **average force × time**, *Ft*.
   The units of impulse are **newton seconds**, Ns.

   $$Ft = mv - mu$$
   (where *v* is the final velocity and *u* is the initial velocity)
   so **impulse = change of momentum**

   Impulse is the area under a force-time graph.

3) So, the **force** of an impact can be **reduced** by **increasing the time** of the impact.

   For example, a toy car with a mass of 1 kg, travelling at 5 ms⁻¹, hits a wall and stops in a time of 0.5 seconds.

   The average force on the car is: $F = \dfrac{mv - mu}{t} = \dfrac{(1 \times 5) - (1 \times 0)}{0.5} = 10 \text{ N}$

   But if the time of impact is doubled to 1 second, the force on the car is halved.

   This is the idea behind car crumple zones which increase the time of an impact to reduce the force on the passengers.

Practice Questions

Q1   Give two examples of conservation of momentum in practice.

Q2   Describe what happens when a tiny object makes an elastic collision with a massive object, and why.

Exam Questions

Q1   A ball of mass 0.6 kg moving at 5 ms⁻¹ collides with a larger stationary ball of mass 2 kg.
     The smaller ball rebounds in the opposite direction at 2.4 ms⁻¹.
     (a) What is the velocity of the larger ball immediately after the collision?   [3 marks]
     (b) Is this an elastic or inelastic collision? Explain your answer.   [3 marks]

Q2   A toy train of mass 0.7 kg, travelling at 0.3 ms⁻¹, collides with a stationary toy carriage of mass 0.4 kg.
     The two toys couple together. What is their new velocity?   [3 marks]

Momentum will never be an endangered species — it's always conserved...

*It seems a bit of a contradiction to say that momentum's always conserved then tell you that impulse is the change in momentum. The difference is that impulse is only talking about the change of momentum of one of the objects, whereas conservation of momentum is talking about the **whole** system.*

Newton's Laws of Motion

You did most of this at GCSE, but that doesn't mean you can just skip over it now. You'll be kicking yourself if you forget this stuff in the exam — easy marks...

Newton's **1st Law** says that a **Force** is Needed to Change Velocity

1) **Newton's 1st law of motion** states the **velocity** of an object will **not change** unless a **resultant force** acts on it.

2) In plain English this means a body will remain at rest or moving in a **straight line** at a **constant speed**, unless acted on by a **resultant force**.

An apple sitting on a table won't go anywhere because the **forces** on it are **balanced**.

reaction (**R**) = weight (mg)

(force of table pushing apple up) (force of gravity pulling apple down)

3) If the forces **aren't balanced**, the **overall resultant force** will cause the body to **accelerate** — if you gave the apple above a shove, there'd be a resultant force acting on it and it would roll off the table. Acceleration can mean a change in **direction**, or **speed**, or both. (See Newton's 2nd law, below.)

Newton's 2nd Law says that Force is the **Rate of Change in Momentum**...

*"The **rate of change of momentum** of an object is **directly proportional** to the **resultant force** which acts on the object."*

so $F = \dfrac{\Delta mv}{\Delta t}$

If mass is constant, this can be written as the well-known equation:

resultant force (F) = mass (m) × acceleration (a)

Learn this — it crops up all over the place in A2 Physics.
And learn what it means too:

1) It says that the **more force** you have acting on a certain mass, the **more acceleration** you get.

2) It says that for a given force the **more mass** you have, the **less acceleration** you get.

REMEMBER:
1) The **resultant force** is the **vector sum** of all the forces.
2) The force is **always** measured in **newtons**. Always.
3) The **mass** is always measured in **kilograms**.
4) **a** is the **acceleration** of the object as a result of F. It's **always** measured in **metres per second per second** (ms^{-2}).
5) The **acceleration** is always in the **same direction** as the **resultant force**.

F = ma is a **Special Case** of Newton's 2nd Law

Newton's 2nd law says that if the **mass** of an object is **constant**, then the **bigger** the **force** acting on it, the **greater** its **acceleration** — i.e. **F = ma**. But, if the **mass** of the object is **changing** — e.g. if it is accelerating at close to the **speed of light** — then you **can't** use **F = ma**.

Don't worry though — **Newton's 2nd law still applies**, it's just that the 'rate of **change of momentum**' bit refers to a **change in mass** and velocity.

Daisy was always being told that she was a special case.

Newton's Laws of Motion

Newton's **3rd Law** *says each Force has an* **Equal**, *Opposite Reaction Force*

There are a few different ways of stating Newton's 3rd law, but the clearest way is:

> **If an object A EXERTS a FORCE on object B, then object B exerts AN EQUAL BUT OPPOSITE FORCE on object A.**

You'll also hear the law as "every action has an equal and opposite reaction". But this confuses people who wrongly think the forces are both applied to the same object. (If that were the case, you'd get a resultant force of zero and nothing would ever move anywhere...)

The two forces actually represent the **same interaction**, just seen from two **different perspectives**:

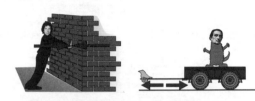

1) If you **push against a wall**, the wall will **push back** against you, **just as hard**. As soon as you stop pushing, so does the wall. Amazing...

2) If you **pull a cart**, whatever force **you exert** on the rope, the rope exerts the **exact opposite** pull on you.

3) When you go **swimming**, you push **back** against the water with your arms and legs, and the water pushes you **forwards** with an equal-sized force.

This looks like Newton's 3rd law...

But it's <u>NOT</u>.

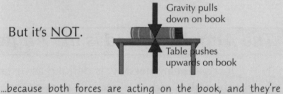

Gravity pulls down on book

Table pushes upwards on book

...because both forces are acting on the book, and they're not of the same type. This is two separate interactions. The forces are equal and opposite, resulting in zero acceleration, so this is showing Newton's 1st law.

Newton's 3rd law applies in **all situations** and to all **types of force**. But the pairs of forces are always the **same type**, e.g. both gravitational or both electrical.

Newton's 3rd law is a consequence of the **conservation of momentum** (page 86). A **resultant force** acting means a change in **mass** or **acceleration** ($F = ma$) — which means a **change in momentum**. Momentum is always **conserved**, so whenever one object exerts a force on another (and changes its momentum) the second object must exert an equal-sized force back onto the first object so that the overall change in momentum is zero.

Practice Questions

Q1 State Newton's 1st, 2nd and 3rd laws of motion, and explain what they mean.

Q2 Give an example of a situation where you couldn't use F = ma. Why wouldn't the equation apply?

Q3 Sketch a force diagram of a book resting on a table to illustrate Newton's 3rd law.

Exam Questions

Q1 A parachutist with a mass of 78 kg jumps out of a plane. As she falls, the resultant force acting on her changes.

    (a) Use Newton's 2nd law to explain why she initially accelerates. [2 marks]

    (b) What is the initial vertical force on the parachutist? Use $g = 9.81$ ms^{-2}. [1 mark]

    (c) After a time, the parachutist reaches terminal velocity and stops accelerating.
        Use Newton's 1st law to explain why the resultant force on the parachutist is zero at this point. [2 marks]

Q2 A boat is moving across a river. The engines provide a force of 500 N at right angles to the flow of the river, and the boat experiences a drag of 100 N in the opposite direction. The force on the boat due to the flow of the river is 300 N. The mass of the boat is 250 kg. Calculate the magnitude of the acceleration of the boat. [4 marks]

Newton's three incredibly important laws of motion...

These equations may not really fill you with a huge amount of excitement (and I hardly blame you if they don't)... but it was pretty fantastic at the time — suddenly people actually understood how forces work, and how they affect motion. I mean arguably it was one of the most important scientific discoveries ever...

Circular Motion

*It's probably worth putting a bookmark in here — this stuff is needed **all over** the place.*

Angles can be Expressed in Radians

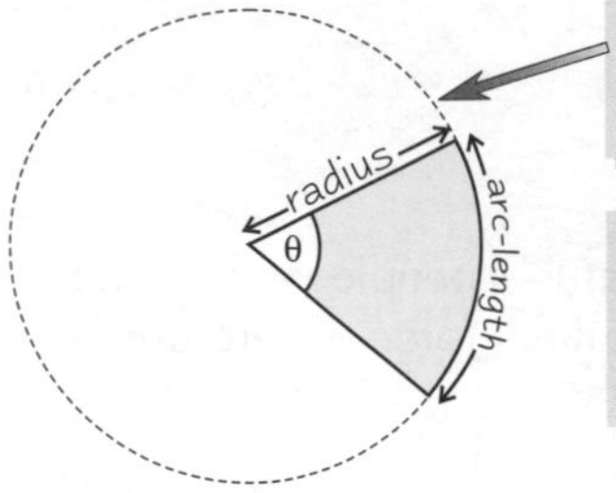

The angle in **radians**, θ, is defined as the **arc-length** divided by the radius of the circle.

For a **complete circle** (360°), the arc-length is just the circumference of the circle ($2\pi r$). Dividing this by the radius (r) gives 2π. So there are 2π radians in a complete circle.

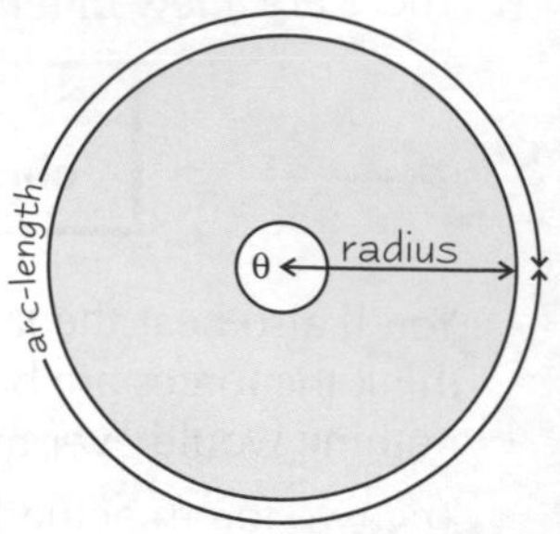

radius   arc-length

θ — radius

Some common angles:

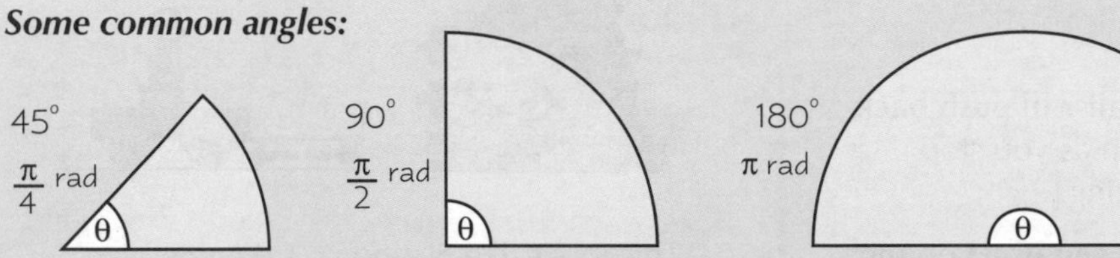

45°
$\dfrac{\pi}{4}$ rad
θ

90°
$\dfrac{\pi}{2}$ rad
θ

180°
π rad
θ

$$\text{angle in radians} = \frac{2\pi}{360} \times \text{angle in degrees}$$

1 radian is about 57°

The **Angular Speed** is the **Angle** an Object Rotates Through **per Second**

1) Just as **linear speed**, v, is defined as distance ÷ time, the **angular speed**, ω, is defined as **angle ÷ time**. The unit is rad s⁻¹ — radians per second.

$$\omega = \frac{\theta}{t}$$

ω = angular speed (rad s⁻¹) — the symbol for angular speed is the little Greek 'omega', not a w.
θ = angle (radians) turned through in a time, t (seconds)

2) The **linear speed**, v, and **angular speed**, ω, of a rotating object are linked by the equation:

$$v = r\omega$$

v = linear speed (ms⁻¹), r = radius of the circle (m),
ω = angular speed (rad s⁻¹)

Example — Beam of Particles in a Cyclotron (a type of particle accelerator)

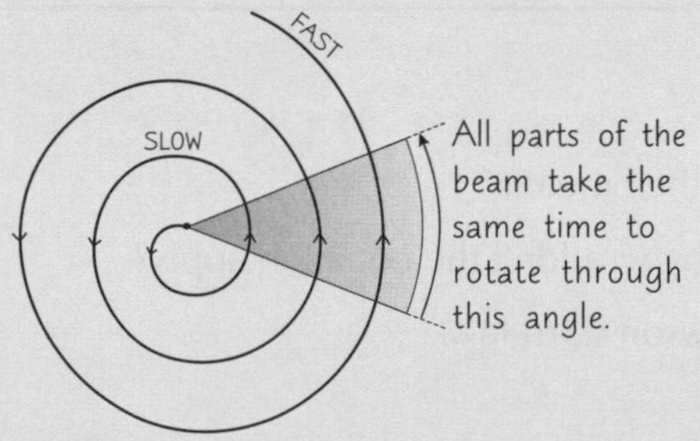

FAST
SLOW
All parts of the beam take the same time to rotate through this angle.

1) Different parts of the particle beam are rotating at **different linear speeds**, v. (The linear speed is sometimes called **tangential velocity**.)

2) But all the parts **rotate** through the **same angle** in the **same time** — so they have the **same angular speed**.

Circular Motion has a **Frequency** and **Period**

1) The frequency, f, is the number of complete **revolutions per second** (rev s⁻¹ or hertz, Hz).

2) The period, T, is the **time taken** for a complete revolution (in seconds).

3) Frequency and period are **linked** by the equation:

$$f = \frac{1}{T}$$

f = frequency in rev s⁻¹, T = period in s

4) For a complete circle, an object turns through 2π radians in a time T, so frequency and period are related to ω by:

$$\omega = 2\pi f \quad \text{and} \quad \omega = \frac{2\pi}{T}$$

f = frequency in rev s⁻¹, T = period in s, ω = angular speed in rad s⁻¹

Circular Motion

Objects Travelling in Circles are **Accelerating** since their **Velocity is Changing**

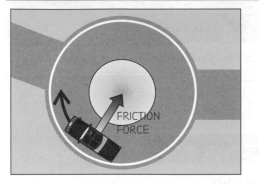

1) Even if the car shown is going at a **constant speed**, its **velocity** is changing since its **direction** is changing.

2) Since acceleration is defined as the **rate of change of velocity**, the car is accelerating even though it isn't going any faster.

3) This acceleration is called the **centripetal acceleration** and is always directed towards the **centre of the circle**.

There are two formulas for centripetal acceleration:

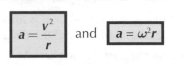

$$a = \frac{v^2}{r} \quad \text{and} \quad a = \omega^2 r$$

a = centripetal acceleration in ms^{-2}
v = linear speed in ms^{-1}
ω = angular speed in rad s^{-1}
r = radius in m

The **Centripetal Acceleration** is produced by a **Centripetal Force**

From Newton's laws, if there's a **centripetal acceleration**, there must be a **centripetal force** acting towards the **centre of the circle**.

Since $F = ma$, the centripetal force must be:

$$F = \frac{mv^2}{r} \quad \text{and} \quad F = m\omega^2 r$$

The centripetal force is what keeps the object moving in a circle — remove the force and the object would fly off at a tangent.

Men cowered from the force of the centripede.

Practice Questions

Q1 How many radians are there in a complete circle?
Q2 How is angular speed defined and what is the relationship between angular speed and linear speed?
Q3 Define the period and frequency of circular motion. What is the relationship between period and angular speed?
Q4 In which direction does the centripetal force act, and what happens when this force is removed?

Exam Questions

Q1 (a) At what angular speed does the Earth orbit the Sun? (1 year = 3.2×10^7 s) [2 marks]

(b) Calculate the Earth's linear speed. (Assume radius of orbit = 1.5×10^{11} m) [2 marks]

(c) Calculate the centripetal force needed to keep the Earth in its orbit. (Mass of Earth = 6.0×10^{24} kg) [2 marks]

(d) What is providing this force? [1 mark]

Q2 A bucket full of water, tied to a rope, is being swung around in a vertical circle (so it is upside down at the top of the swing). The radius of the circle is 1 m.

(a) By considering the acceleration due to gravity at the top of the swing, what is the minimum frequency with which the bucket can be swung without any water falling out? [3 marks]

(b) The bucket is now swung with a constant angular speed of 5 rad s^{-1}. What will be the tension in the rope when the bucket is at the top of the swing if the total mass of the bucket and water is 10 kg? [2 marks]

I'm spinnin' around, move out of my way...

*"Centripetal" just means "centre-seeking". The centripetal force is what actually causes circular motion. What you **feel** when you're spinning, though, is the reaction (centrifugal) force. Don't get the two mixed up.*

Gravitational Fields

*Gravity's all about masses **attracting** each other. If the Earth didn't have a **gravitational field,** apples wouldn't fall to the ground and you'd probably be floating off into space instead of sitting here reading this page...*

Masses in a *Gravitational Field* Experience a *Force of Attraction*

1) Any object with mass will **experience an attractive force** if you put it in the **gravitational field** of another object.

2) Only objects with a **large** mass, such as stars and planets, have a significant effect. E.g. the gravitational fields of the **Moon** and the **Sun** are noticeable here on Earth — they're the main cause of our **tides**.

You can *Calculate Forces* Using *Newton's Law of Gravitation*

1) The **force** experienced by an object in a gravitational field is always **attractive**. It's a **vector** which depends on the **masses** involved and the **distance** between them.

2) The diagram shows the force acting on mass **m** due to mass **M**. (The force on **M** due to **m** is equal but in the opposite direction.)

3) **M** and **m** are uniform spheres, which behave as **point masses** — as if all their mass is concentrated at the centre.

4) It's easy to work out the **force** experienced by a **point mass** in a **gravitational field** — you just put the numbers into this **equation**:

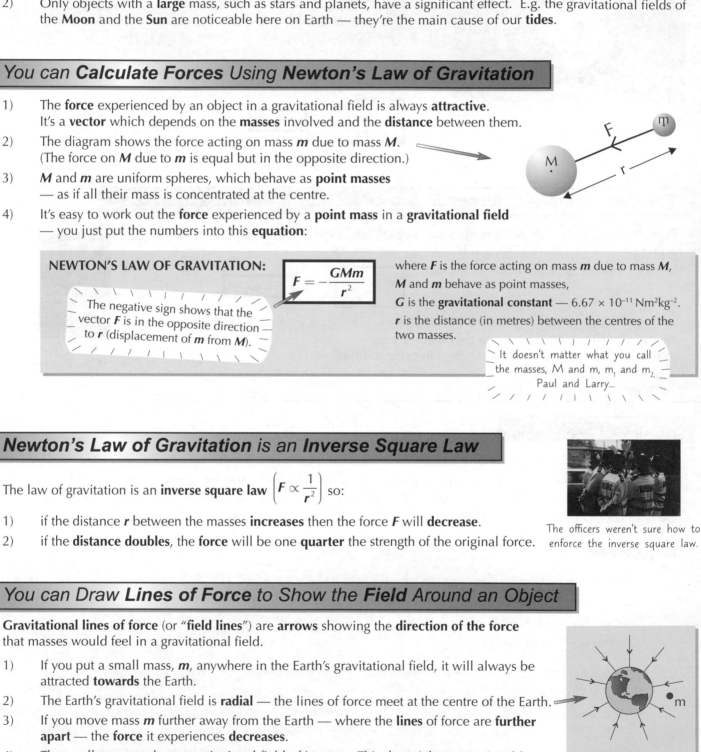

NEWTON'S LAW OF GRAVITATION:

$$F = -\frac{GMm}{r^2}$$

The negative sign shows that the vector **F** is in the opposite direction to **r** (displacement of **m** from **M**).

where **F** is the force acting on mass **m** due to mass **M**,
M and **m** behave as point masses,
G is the **gravitational constant** — 6.67×10^{-11} Nm^2kg^{-2}.
r is the distance (in metres) between the centres of the two masses.

It doesn't matter what you call the masses, M and m, m_1 and m_2, Paul and Larry...

Newton's Law of Gravitation *is an* Inverse Square Law

The law of gravitation is an **inverse square law** $\left(F \propto \dfrac{1}{r^2}\right)$ so:

1) if the distance **r** between the masses **increases** then the force **F** will **decrease**.

2) if the **distance doubles**, the **force** will be one **quarter** the strength of the original force.

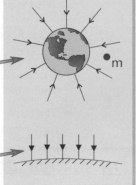

The officers weren't sure how to enforce the inverse square law.

You can Draw *Lines of Force* to Show the *Field* Around an Object

Gravitational lines of force (or "field lines") are **arrows** showing the **direction of the force** that masses would feel in a gravitational field.

1) If you put a small mass, **m**, anywhere in the Earth's gravitational field, it will always be attracted **towards** the Earth.

2) The Earth's gravitational field is **radial** — the lines of force meet at the centre of the Earth.

3) If you move mass **m** further away from the Earth — where the **lines** of force are **further apart** — the **force** it experiences **decreases**.

4) The small mass, **m**, has a gravitational field of its own. This doesn't have a noticeable effect on the Earth though, because the Earth is so much **more massive**.

5) Close to the Earth's surface, the gravitational field is (almost) uniform — so the **field lines** are (almost) **parallel**. You can usually **assume** that the field is perfectly uniform.

Gravitational Fields

The **Field Strength** is the **Force per Unit Mass**

Gravitational field strength, g, is the **force per unit mass**. Its value depends on **where you are** in the field. There's a really simple equation for working it out:

$$g = \frac{F}{m}$$

g has units of newtons per kilogram (Nkg^{-1})

Murray loved showing off his field strength.

1) F is the force experienced by a mass m when it's placed in the gravitational field. Divide F by m and you get the **force per unit mass**.

2) g is a **vector** quantity, always pointing towards the centre of the mass whose field you're describing.

3)   Since the gravitational field is almost uniform at the Earth's surface, you can assume g is a constant.

4) g is just the **acceleration** of a mass in a gravitational field. It's often called the **acceleration due to gravity**.

> The **value** of g at the **Earth's surface** is approximately **9.81** ms^{-2} (or 9.81 Nkg^{-1}).

In a **Radial Field**, g is **Inversely Proportional** to r^2

Point masses have **radial** gravitational fields. The value of g depends on the distance r from the centre of the point mass M...

$$g = -\frac{GM}{r^2}$$

The graph shows how g varies for the Earth. R_E is the Earth's radius

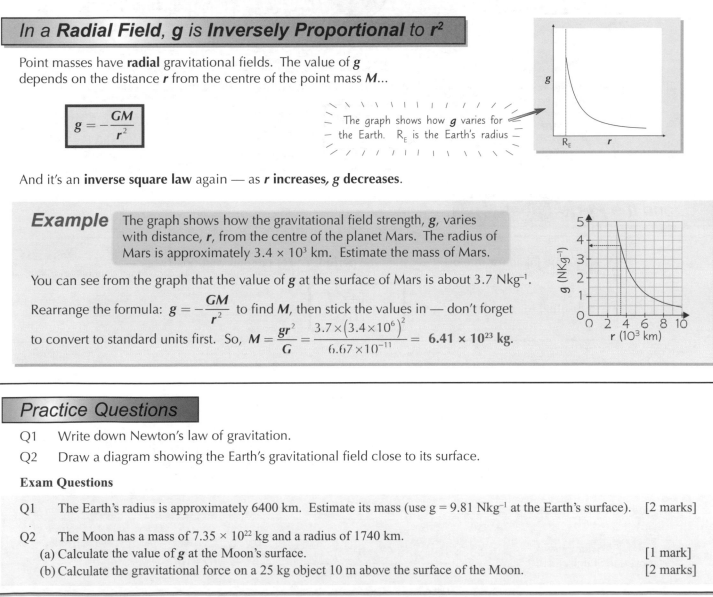

And it's an **inverse square law** again — as r **increases**, g **decreases**.

> **Example**   The graph shows how the gravitational field strength, g, varies with distance, r, from the centre of the planet Mars. The radius of Mars is approximately 3.4×10^3 km. Estimate the mass of Mars.
>
> You can see from the graph that the value of g at the surface of Mars is about 3.7 Nkg^{-1}.
>
> Rearrange the formula: $g = -\frac{GM}{r^2}$ to find M, then stick the values in — don't forget to convert to standard units first. So, $M = \frac{gr^2}{G} = \frac{3.7 \times (3.4 \times 10^6)^2}{6.67 \times 10^{-11}} = \mathbf{6.41 \times 10^{23}}$ **kg.**

Practice Questions

Q1   Write down Newton's law of gravitation.

Q2   Draw a diagram showing the Earth's gravitational field close to its surface.

Exam Questions

Q1   The Earth's radius is approximately 6400 km. Estimate its mass (use $g = 9.81$ Nkg^{-1} at the Earth's surface).   [2 marks]

Q2   The Moon has a mass of 7.35×10^{22} kg and a radius of 1740 km.
    (a) Calculate the value of g at the Moon's surface.   [1 mark]
    (b) Calculate the gravitational force on a 25 kg object 10 m above the surface of the Moon.   [2 marks]

If you're really stuck, put 'Inverse Square Law'...

Clever chap, Newton, but famously tetchy. He got into fights with other physicists, mainly over planetary motion and calculus... the usual playground squabbles. Then he spent the rest of his life trying to turn scrap metal into gold. Weird.

Motion of Masses in Gravitational Fields

*Planets just go round and round in circles. Well, **ellipses** really, but I won't tell if you don't...*

Planets are Satellites which Orbit the Sun

1) A **satellite** is just any **smaller mass** which **orbits** a **much larger mass** — the **Moon** is a satellite of the Earth.

2) In our Solar System, the planets have **nearly circular orbits**... so you can use the **equations of circular motion**.

The Speed of an Orbit depends on its Radius and the Mass of the Larger Body

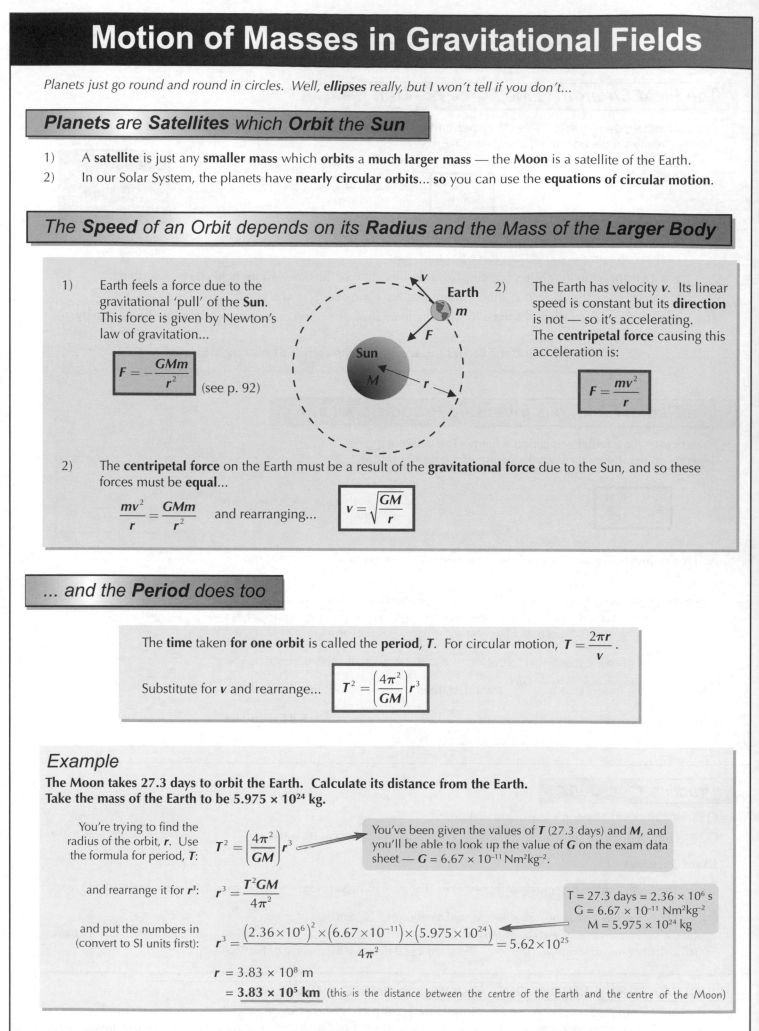

1) Earth feels a force due to the gravitational 'pull' of the **Sun**. This force is given by Newton's law of gravitation...

$$F = -\frac{GMm}{r^2}$$ (see p. 92)

2) The Earth has velocity **v**. Its linear speed is constant but its **direction** is not — so it's accelerating. The **centripetal force** causing this acceleration is:

$$F = \frac{mv^2}{r}$$

2) The **centripetal force** on the Earth must be a result of the **gravitational force** due to the Sun, and so these forces must be **equal**...

$$\frac{mv^2}{r} = \frac{GMm}{r^2}$$ and rearranging...

$$v = \sqrt{\frac{GM}{r}}$$

... and the Period does too

The **time** taken **for one orbit** is called the **period**, **T**. For circular motion, $T = \frac{2\pi r}{v}$.

Substitute for **v** and rearrange...

$$T^2 = \left(\frac{4\pi^2}{GM}\right)r^3$$

Example

The Moon takes 27.3 days to orbit the Earth. Calculate its distance from the Earth.
Take the mass of the Earth to be 5.975 × 10²⁴ kg.

You're trying to find the radius of the orbit, **r**. Use the formula for period, **T**:

$$T^2 = \left(\frac{4\pi^2}{GM}\right)r^3$$

You've been given the values of **T** (27.3 days) and **M**, and you'll be able to look up the value of **G** on the exam data sheet — **G** = 6.67 × 10⁻¹¹ Nm²kg⁻².

and rearrange it for **r³**:

$$r^3 = \frac{T^2GM}{4\pi^2}$$

T = 27.3 days = 2.36 × 10⁶ s
G = 6.67 × 10⁻¹¹ Nm²kg⁻²
M = 5.975 × 10²⁴ kg

and put the numbers in (convert to SI units first):

$$r^3 = \frac{\left(2.36\times10^6\right)^2\times\left(6.67\times10^{-11}\right)\times\left(5.975\times10^{24}\right)}{4\pi^2} = 5.62\times10^{25}$$

$$r = 3.83 \times 10^8 \text{ m}$$

$$= \underline{\textbf{3.83} \times \textbf{10}^5 \textbf{ km}}$$ (this is the distance between the centre of the Earth and the centre of the Moon)

Motion of Masses in Gravitational Fields

Geostationary Satellites Orbit the Earth once in 24 hours

1) Geostationary satellites orbit directly over the **equator** and are **always above the same point** on Earth.
2) A geostationary satellite travels at the **same angular speed as the Earth** turns below it.
3) These satellites are really useful for sending TV and telephone signals — the satellite is **stationary** relative to a certain point on the **Earth**, so you don't have to alter the angle of your receiver (or transmitter) to keep up.
4) Their orbit takes exactly **one day**.

Kepler's Laws are about the Motion of Planets in the Solar System

Kepler came up with these three laws around 1600, about 80 years before Newton developed his law of gravitation:

1) Each planet moves in an **ellipse** around the Sun (a circle is just a special kind of ellipse).
2) A line joining the Sun to a planet will sweep out **equal areas in equal times**.
(Don't worry, you don't need to know about that one.)
3) The **period** of the orbit and the **mean distance** between the Sun and the planet are related by **Kepler's third law**:

$$T^2 \propto r^3$$

For circular motion, where r = radius of the orbit, $\dfrac{r^3}{T^2} = \text{constant} = \dfrac{GM}{4\pi^2}$.

It's either gravity or a giant white rabbit that makes the Earth orbit the Sun. I know which one I believe...

Example   The diagram shows the orbits of two of Jupiter's moons, Io and Europa. The moon Io completes one orbit of Jupiter in 42.5 hours. Estimate the orbital period of Europa to the nearest hour, assuming both orbits are circular.

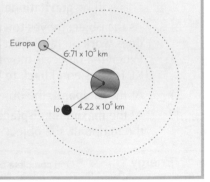

Using Kepler's third law: $\dfrac{T_{Io}^{\,2}}{r_{Io}^{\,3}} = \dfrac{T_{Europa}^{\,2}}{r_{Europa}^{\,3}}$,

so $T_{Europa} = \sqrt{\dfrac{T_{Io}^{\,2}\, r_{Europa}^{\,3}}{r_{Io}^{\,3}}} = \sqrt{\dfrac{(42.5)^2 \times (6.71\times10^5)^3}{(4.22\times10^5)^3}} = 85$ hours (to the nearest hour).

Practice Questions

Q1 Derive an expression for the radius of the orbit of a planet around the Sun, in terms of the period of its orbit.

Q2 The International Space Station orbits the Earth with velocity v. If another vehicle docks with it, increasing its mass, what difference, if any, does this make to the speed or radius of the orbit?

Q3 Would a geostationary satellite be useful for making observations for weather forecasts? Give reasons.

Exam Questions

(Use $G = 6.67 \times 10^{-11}\,Nm^2kg^{-2}$, mass of Earth = 5.98×10^{24} kg, radius of Earth = 6400 km)

Q1 (a) A satellite orbits 200 km above the Earth's surface. Calculate the period of the satellite's orbit. [2 marks]
    (b) What is the linear speed of the satellite? [1 mark]

Q2 At what height above the Earth's surface would a geostationary satellite orbit? [3 marks]

Q3 The Sun has a mass of 2.0×10^{30} kg, but loses mass at a rate of around 6×10^9 kgs^{-1}. Discuss whether this will have had any significant effect on the Earth's orbit over the past 50 000 years. [2 marks]

No fluffy bunnies were harmed in the making of these pages...

Kepler is sometimes proclaimed as the first science fiction writer. He wrote a tale about a fantastic trip to the Moon, where the book narrator's mum asks a demon the secret of space travel, to boldly go where — oh wait, different story. Unfortunately Kepler's book might have sparked the actual witchhunt on Kepler's mum, whoops-a-daisy...

Simple Harmonic Motion

Something simple at last, I like the sound of this. And colourful graphs too — you're in for a treat here.

SHM is Defined in terms of Acceleration and Displacement

1) An object moving with **simple harmonic motion** (SHM) **oscillates** to and fro, either side of a **midpoint**.

2) **Pendulums** and **mass-spring systems** (a mass hanging on a spring that's free to move up and down) are two examples.

3) The distance of the object from the midpoint is called its **displacement**.

4) There is always a **restoring force** pulling or pushing the object back **towards** the **midpoint**.

5) The **size** of the **restoring force** depends on the **displacement**, and the force makes the object **accelerate** towards the midpoint:

> **SHM:** an oscillation in which the **acceleration** of an object is **directly proportional** to its **displacement** from the **midpoint**, and is directed **towards the midpoint**.

The Restoring Force makes the Object Exchange PE and KE

1) The **type** of **potential energy** (PE) depends on **what it is** that's providing the **restoring force**. This will be **gravitational PE** for pendulums and **elastic PE** (elastic stored energy) for masses on springs.

2) As the object moves **towards the midpoint**, the restoring force **does work** on the object and so **transfers** some PE to KE. When the object is moving **away from the midpoint**, all that KE is transferred **back to PE** again.

3) At the **midpoint**, the object's **PE** is **zero** and its **KE** is **maximum**.

4) At the **maximum displacement** (the **amplitude**) on both sides of the midpoint, the object's **KE** is **zero** and its **PE** is **maximum**.

5) The **sum** of the **potential** and **kinetic** energy is called the **mechanical energy** and **stays constant** (as long as the motion isn't damped — see p. 98-99).

6) The **energy transfer** for one complete cycle of oscillation (see graph) is: PE to KE to PE to KE to PE ... and then the process repeats...

You can Draw Graphs to Show Displacement, Velocity and Acceleration

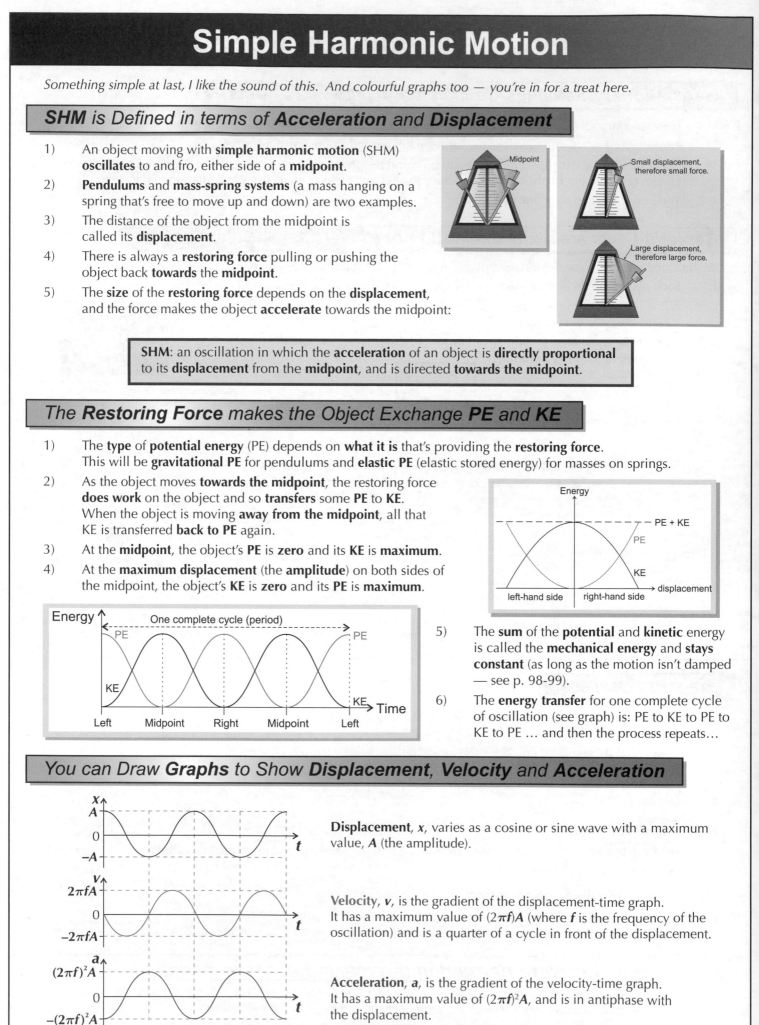

Displacement, x, varies as a cosine or sine wave with a maximum value, A (the amplitude).

Velocity, v, is the gradient of the displacement-time graph. It has a maximum value of $(2\pi f)A$ (where f is the frequency of the oscillation) and is a quarter of a cycle in front of the displacement.

Acceleration, a, is the gradient of the velocity-time graph. It has a maximum value of $(2\pi f)^2 A$, and is in antiphase with the displacement.

Simple Harmonic Motion

The *Frequency* and *Period* don't depend on the *Amplitude*

1) From **maximum positive displacement** (e.g. maximum displacement to the right) to **maximum negative displacement** (e.g. maximum displacement to the left) and **back again** is called a **cycle** of oscillation.

2) The **frequency**, *f*, of the SHM is the number of cycles per second (measured in Hz).

3) The **period**, *T*, is the **time** taken for a complete cycle (in seconds).

> In SHM, the **frequency** and **period** are independent of the **amplitude** (i.e. constant for a given oscillation). So a pendulum clock will keep ticking in regular time intervals even if its swing becomes very small.

Learn the SHM Equations

I know it looks like there are loads of complicated equations to learn here, but don't panic — it's not that bad really. You'll be given these formulas in the exam, so just make sure you know what they mean and how to use them.

1) According to the definition of SHM, the **acceleration**, *a*, is directly proportional to the **displacement**, *x*. The **constant of proportionality** depends on the **frequency**, and the acceleration is always in the **opposite direction** from the displacement (so there's a minus sign in the equation).

$$a = -(2\pi f)^2 x$$

2) The **velocity** is **positive** if the object's moving **away** from the **midpoint**, and **negative** if it's moving **towards** the midpoint. It reaches a maximum, v_{max}, when the object is at the midpoint of its oscillation ($x = 0$).

Max velocity: $v_{max} = 2\pi f A$

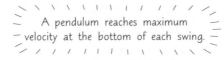

A pendulum reaches maximum velocity at the bottom of each swing.

3) The **displacement** varies with time according to one of two equations, depending on **where** the object was when the timing was started. This determines whether the displacement-time graph is a cosine or sine wave.

For someone starting a stopwatch with a pendulum at **maximum displacement**:

$$x = A\cos(2\pi f t)$$

For someone starting a stopwatch as the pendulum swings through **the midpoint**:

$$x = A\sin(2\pi f t)$$

Practice Questions

Q1 Sketch a graph of how the velocity of an object oscillating with SHM varies with time.

Q2 What is the special relationship between the acceleration and the displacement in SHM?

Q3 Given the amplitude and the frequency, how would you work out the maximum acceleration?

Exam Questions

Q1 (a) Define *simple harmonic motion*. [2 marks]
(b) Explain why the motion of a ball bouncing off the ground is not SHM. [1 mark]

Q2 A pendulum is pulled a distance 0.05 m from its midpoint and released.
It oscillates with simple harmonic motion with a frequency of 1.5 Hz. Calculate:
(a) its maximum velocity [1 mark]
(b) its displacement 0.1 s after it is released [2 marks]
(c) the time it takes to fall to 0.01 m from the midpoint after it is released [2 marks]

"Simple" harmonic motion — hmmm, I'm not convinced...

The basic concept of SHM is simple enough (no pun intended). Make sure you can remember the shapes of all the graphs on page 96 and the equations from this page, then just get as much practice at using the equations as you can.

Free and Forced Vibrations

Resonance... hmm... tricky little beast. Remember the Millennium Bridge, that standard-bearer of British engineering? The wibbles and wobbles were caused by resonance. How was it sorted out? By damping, which is coming up too.

Free Vibrations — No Transfer of Energy To or From the Surroundings

1) If you stretch and release a mass on a spring, it oscillates at its **natural frequency**.

2) If **no energy's transferred** to or from the surroundings, it will **keep** oscillating with the **same amplitude forever**.

3) In practice this **never happens**, but a spring vibrating in air is called a **free vibration** anyway.

Forced Vibrations happen when there's an External Driving Force

1) A system can be **forced** to vibrate by a periodic **external force**.

2) The frequency of this force is called the **driving frequency**.

Resonance happens when Driving Frequency = Natural Frequency

When the **driving frequency** approaches the **natural frequency**, the system gains more and more energy from the driving force and so vibrates with a **rapidly increasing amplitude**. When this happens the system is **resonating**.

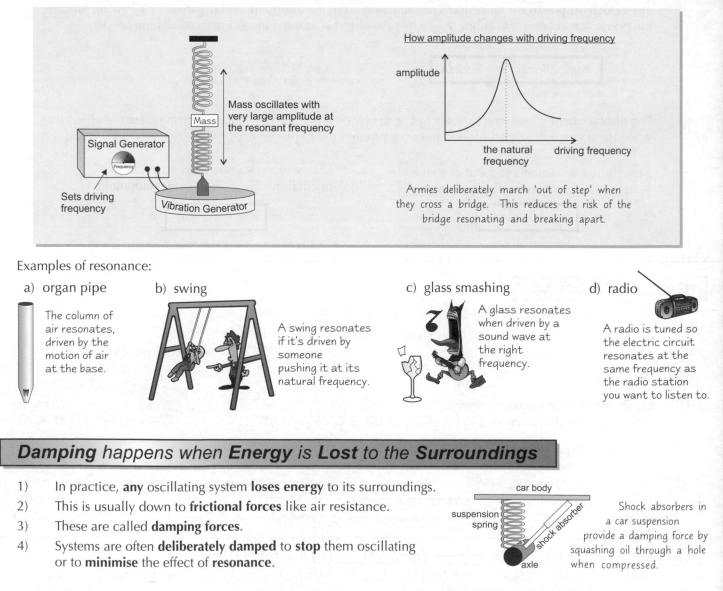

Mass oscillates with very large amplitude at the resonant frequency

Signal Generator — Sets driving frequency

Vibration Generator

How amplitude changes with driving frequency

amplitude

the natural frequency   driving frequency

Armies deliberately march 'out of step' when they cross a bridge. This reduces the risk of the bridge resonating and breaking apart.

Examples of resonance:

a) organ pipe
The column of air resonates, driven by the motion of air at the base.

b) swing
A swing resonates if it's driven by someone pushing it at its natural frequency.

c) glass smashing
A glass resonates when driven by a sound wave at the right frequency.

d) radio
A radio is tuned so the electric circuit resonates at the same frequency as the radio station you want to listen to.

Damping happens when Energy is Lost to the Surroundings

1) In practice, **any** oscillating system **loses energy** to its surroundings.

2) This is usually down to **frictional forces** like air resistance.

3) These are called **damping forces**.

4) Systems are often **deliberately damped** to **stop** them oscillating or to **minimise** the effect of **resonance**.

car body

suspension spring

shock absorber

axle

Shock absorbers in a car suspension provide a damping force by squashing oil through a hole when compressed.

Free and Forced Vibrations

Different Amounts of Damping have Different Effects

1) The **degree** of damping can vary from **light** damping (where the damping force is small) to **overdamping**.

2) Damping **reduces** the **amplitude** of the oscillation over time. The **heavier** the damping, the **quicker** the amplitude is reduced to zero.

3) **Critical damping** reduces the amplitude (i.e. stops the system oscillating) in the **shortest possible time**.

4) Car **suspension systems** and moving coil **meters** are critically damped so that they **don't oscillate** but return to equilibrium as quickly as possible.

5) Systems with **even heavier damping** are **overdamped**. They take **longer** to return to equilibrium than a critically damped system.

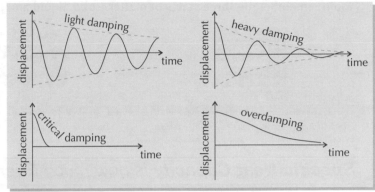

6) **Plastic deformation** of ductile materials **reduces** the **amplitude** of oscillations in the same way as damping. As the material changes shape, it **absorbs energy**, so the oscillation will become smaller.

Damping Affects Resonance too

1) **Lightly damped** systems have a **very sharp** resonance peak. Their amplitude only increases dramatically when the **driving frequency** is **very close** to the **natural frequency**.

2) **Heavily damped** systems have a **flatter response**. Their amplitude doesn't increase very much near the natural frequency and they aren't as **sensitive** to the driving frequency.

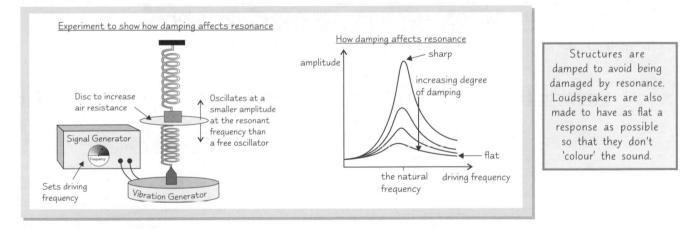

Structures are damped to avoid being damaged by resonance. Loudspeakers are also made to have as flat a response as possible so that they don't 'colour' the sound.

Practice Questions

Q1 What is a free vibration? What is a forced vibration?

Q2 Draw diagrams to show how a damped system oscillates with time when the system is lightly damped and when the system is critically damped.

Exam Questions

Q1 (a) What is resonance? [2 marks]
(b) Draw a diagram to show how the amplitude of a lightly damped system varies with driving frequency. [2 marks]
(c) On the same diagram, show how the amplitude of the system varies with driving frequency when it is heavily damped. [1 mark]

Q2 (a) What is critical damping? [1 mark]
(b) Describe a situation where critical damping is used. [1 mark]

A2 Physics — it can really put a damper on your social life...

Resonance can be really useful (radios, oboes, swings — yay) or very, <u>very</u> bad...

Solids, Liquids and Gases

You need energy to heat something up, and to change its state. Everything comes down to energy. Pretty much always.

The **Three States of Matter** are Defined by the **Arrangement of Particles**

You'll remember the **three states of matter** (**solid**, **liquid** and **gas**) from GCSE — but here's a quick recap just in case.

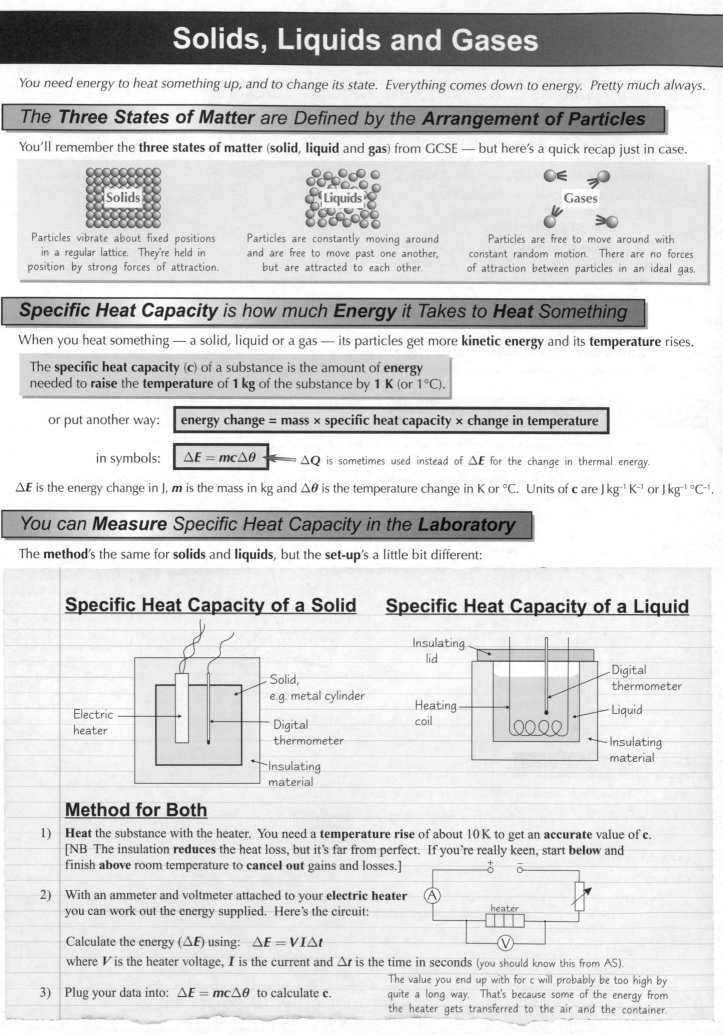

Solids

Particles vibrate about fixed positions in a regular lattice. They're held in position by strong forces of attraction.

Liquids

Particles are constantly moving around and are free to move past one another, but are attracted to each other.

Gases

Particles are free to move around with constant random motion. There are no forces of attraction between particles in an ideal gas.

Specific Heat Capacity is how much Energy it Takes to Heat Something

When you heat something — a solid, liquid or a gas — its particles get more **kinetic energy** and its **temperature** rises.

The **specific heat capacity** (**c**) of a substance is the amount of **energy** needed to **raise** the **temperature** of **1 kg** of the substance by **1 K** (or **1°C**).

or put another way: **energy change = mass × specific heat capacity × change in temperature**

in symbols: $\Delta E = mc\Delta\theta$ ←── ΔQ is sometimes used instead of ΔE for the change in thermal energy.

ΔE is the energy change in J, **m** is the mass in kg and $\Delta\theta$ is the temperature change in K or °C. Units of **c** are $J\,kg^{-1}\,K^{-1}$ or $J\,kg^{-1}\,°C^{-1}$.

You can Measure Specific Heat Capacity in the Laboratory

The **method**'s the same for **solids** and **liquids**, but the **set-up**'s a little bit different:

Specific Heat Capacity of a Solid   Specific Heat Capacity of a Liquid

Electric heater — Solid, e.g. metal cylinder — Digital thermometer — Insulating material

Insulating lid — Digital thermometer — Heating coil — Liquid — Insulating material

Method for Both

1) **Heat** the substance with the heater. You need a **temperature rise** of about 10 K to get an **accurate** value of **c**. [NB The insulation **reduces** the heat loss, but it's far from perfect. If you're really keen, start **below** and finish **above** room temperature to **cancel out** gains and losses.]

2) With an ammeter and voltmeter attached to your **electric heater** you can work out the energy supplied. Here's the circuit:

Calculate the energy (ΔE) using: $\Delta E = VI\Delta t$

where V is the heater voltage, I is the current and Δt is the time in seconds (you should know this from AS).

3) Plug your data into: $\Delta E = mc\Delta\theta$ to calculate **c**.

The value you end up with for c will probably be too high by quite a long way. That's because some of the energy from the heater gets transferred to the air and the container.

Solids, Liquids and Gases

Use $\Delta E = mc\Delta\theta$ to Calculate Specific Heat Capacity

Example You heat 0.25 kg of water from 12.1 °C to 22.9 °C with an electric immersion heater. The heater has a voltage of 11.2 V and a current of 5.3 A, and is switched on for 205 s.

Electrical energy supplied $= V I \Delta t = 11.2 \times 5.3 \times 205 = 12\,170$ J

Temperature rise $= 22.9 - 12.1 = 10.8\,°C = 10.8$ K

So $c = \dfrac{12170}{0.25 \times 10.8} = 4510$ Jkg⁻¹K⁻¹

The actual value for water is 4180 J kg⁻¹ K⁻¹. This result's too big, because ΔE is bigger than it should be (like I said before).

Specific Latent Heat is Defined as the Latent Heat per kg

To **melt** a **solid**, you need to **break the bonds** that hold the particles in place. The **energy** needed for this is called the **latent heat of fusion**. Similarly, when you **boil or evaporate a liquid, energy is needed** to **pull the particles apart** completely. This is the **latent heat of vaporisation**.

The **larger** the **mass** of the substance, the **more energy** it takes to **change** its **state**. That's why the **specific latent heat** is defined per kg:

The **specific latent heat** (l) of **fusion** or **vaporisation** is the quantity of **thermal energy** required to **change the state** of **1 kg** of a substance.

which gives: **energy change = specific latent heat × mass of substance changed**

or in symbols: $\Delta E = ml$

You'll usually see the latent heat of vaporisation written l_v and the latent heat of fusion written l_f.

Where ΔE is the energy change in J and m is the mass in kg. The units of l are J kg⁻¹.

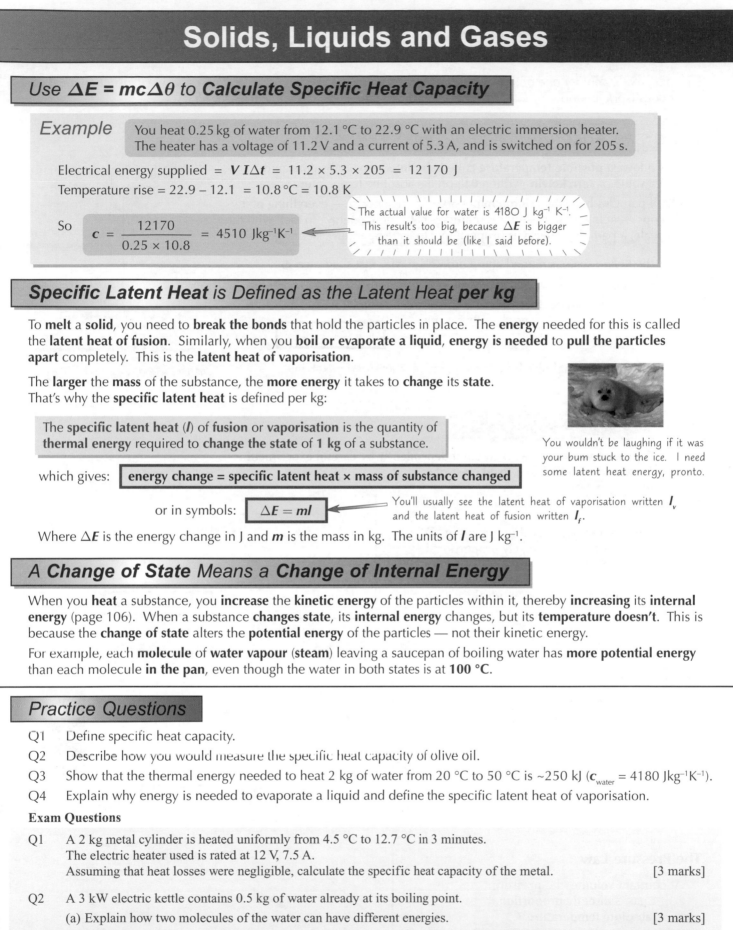

You wouldn't be laughing if it was your bum stuck to the ice. I need some latent heat energy, pronto.

A Change of State Means a Change of Internal Energy

When you **heat** a substance, you **increase** the **kinetic energy** of the particles within it, thereby **increasing** its **internal energy** (page 106). When a substance **changes state**, its **internal energy** changes, but its **temperature doesn't**. This is because the **change of state** alters the **potential energy** of the particles — not their kinetic energy.

For example, each **molecule** of **water vapour** (**steam**) leaving a saucepan of boiling water has **more potential energy** than each molecule **in the pan**, even though the water in both states is at **100 °C**.

Practice Questions

Q1   Define specific heat capacity.

Q2   Describe how you would measure the specific heat capacity of olive oil.

Q3   Show that the thermal energy needed to heat 2 kg of water from 20 °C to 50 °C is ~250 kJ ($c_{water} = 4180$ Jkg⁻¹K⁻¹).

Q4   Explain why energy is needed to evaporate a liquid and define the specific latent heat of vaporisation.

Exam Questions

Q1   A 2 kg metal cylinder is heated uniformly from 4.5 °C to 12.7 °C in 3 minutes. The electric heater used is rated at 12 V, 7.5 A. Assuming that heat losses were negligible, calculate the specific heat capacity of the metal. [3 marks]

Q2   A 3 kW electric kettle contains 0.5 kg of water already at its boiling point.

(a) Explain how two molecules of the water can have different energies. [3 marks]

(b) Neglecting heat losses, how long will it take to boil dry? (l_v (water) $= 2.26 \times 10^6$ J kg⁻¹) [3 marks]

My specific eat capacity — 24 pies...

*This stuff's a bit dull, but hey... make sure you're comfortable using those equations. Interesting(ish) fact for the day — it's the **huge** difference in specific heat capacity between the land and the sea that causes the monsoon in Asia. So there.*

Ideal Gases

*Aaahh... great... another one of those 'our equation doesn't work properly with **real gases**, so we'll invent an **ideal** gas that it **does work** for and they'll think we're dead clever' situations. Hmm. Physicists, eh...*

There's an **Absolute Scale** of Temperature

There is a **lowest possible temperature** called **absolute zero***. Absolute zero is given a value of **zero kelvin**, written **0 K**, on the absolute temperature scale.

At **0 K** all particles have the **minimum** possible **kinetic energy** — everything pretty much stops — at higher temperatures, particles have more energy. In fact, with the **Kelvin scale**, a particle's **energy** is **proportional** to its **temperature** (see page 107).

1) The Kelvin scale is named after Lord Kelvin who first suggested it.

2) A change of **1 K** equals a change of **1 °C**.

3) To change from degrees Celsius into kelvin you **add 273.15** (or subtract 273.15 to go the other way).

$$K = C + 273.15$$

All equations in **thermal physics** use temperatures measured in kelvin.

Equivalent temperatures
373 K — 100 °C
273 K — 0 °C
0 K — −273 °C

**It's true. −273.15 °C is the lowest temperature theoretically possible. Weird, huh. You'd kinda think there wouldn't be a minimum, but there is.*

There are **Three Gas Laws**

The three gas laws were each worked out **independently** by **careful experiment**. Each of the gas laws applies to a **fixed mass** of gas.

Boyle's Law

At a **constant temperature** the **pressure p** and **volume V** of a gas are **inversely proportional**.

A (theoretical) gas that obeys Boyle's law at all temperatures is called an **ideal gas**.

The higher the temperature of the gas, the further the curve is from the origin.

$$pV = \text{constant}$$

Charles' Law

At constant **pressure**, the **volume V** of a gas is **directly proportional** to its **absolute temperature T**.

Ideal gases obey this law and the pressure law as well.

For any ideal gas, the line meets the temperature axis at −273.15 °C — that is, absolute zero.

$$V/T = \text{constant}$$

'Ello, 'ello...

The Pressure Law

At constant **volume**, the **pressure p** of a gas is **directly proportional** to its **absolute temperature T**.

If you'd plotted these graphs in kelvin, they'd both have gone through the origin.

$$p/T = \text{constant}$$

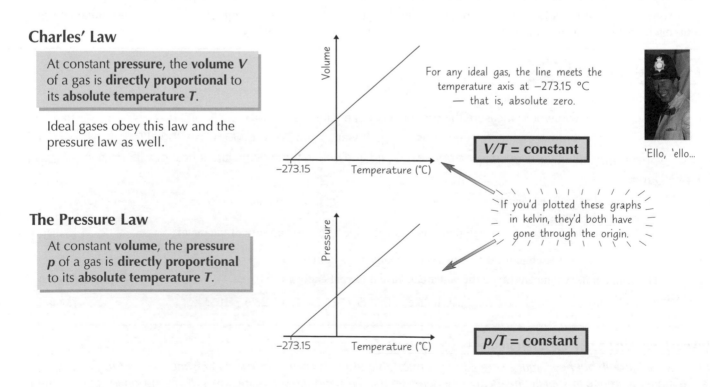

Ideal Gases

If you **Combine** All Three you get the **Ideal Gas Equation**

Combining all three gas laws gives the equation: $p\dfrac{V}{T} = $ **constant**

1) The constant in the equation depends on the amount of gas used. ⟵ (Pretty obvious... if you have more gas it takes up more space.)
The amount of **gas** can be **measured** in **moles**, *n*.

2) The constant then becomes **nR**, where **R** is called the **molar gas constant**.
Its value is 8.31 J mol⁻¹ K⁻¹.

3) Plugging this into the equation gives: $p\dfrac{V}{T} = nR$ or rearranging, **pV = nRT — the ideal gas equation**

This equation works well (i.e., real gases approximate an ideal gas) for gases at **low pressure** and fairly **high temperatures**.

Boltzmann's Constant k is like a **Gas Constant** for **One Particle** of Gas

One mole of any **gas** contains the same number of particles.
This number is called **Avogadro's constant** and has the symbol N_A. The value of N_A is **6.02 × 10²³ particles per mole**.

1) The **number of particles** in a **mass of gas** is given by the **number of moles**, *n*, multiplied by **Avogadro's constant**.
So the number of particles, $N = nN_A$.

2) **Boltzmann's constant**, *k*, is equivalent to R/N_A — you can think of Boltzmann's constant as the **gas constant** for **one particle of gas**, while **R** is the gas constant for **one mole of gas**.

3) The value of Boltzmann's constant is **1.38 × 10⁻²³ JK⁻¹**.

4) If you combine $N = nN_A$ and $k = R/N_A$ you'll see that $Nk = nR$
— which can be substituted into the ideal gas equation: ⟶ **pV = NkT — the equation of state**

The equation **pV = NkT** is called the equation of state of an ideal gas.

Practice Questions

Q1 State Boyle's law, Charles' law and the pressure law.

Q2 What is the ideal gas equation?

Q3 The pressure of a gas is 100 000 Pa and its temperature is 27 °C. The gas is heated — its volume stays fixed but the pressure rises to 150 000 Pa. Show that its new temperature is 177 °C.

Q4 What is the equation of state of an ideal gas?

Exam Questions

Q1 The mass of one mole of nitrogen gas is 0.028 kg. R = 8.31 J mol⁻¹ K⁻¹.

(a) A flask contains 0.014 kg of nitrogen gas.

i) How many moles of nitrogen gas are in the flask? [1 mark]

ii) How many nitrogen molecules are in the flask? [1 mark]

(b) The flask has a volume of 0.01 m³ and is at a temperature of 27 °C. What is the pressure inside it? [2 marks]

(c) What would happen to the pressure if the number of molecules of nitrogen in the flask was halved? [1 mark]

Q2 A large helium balloon has a volume of 10 m³ at ground level.
The temperature of the gas in the balloon is 293 K and the pressure is 1 × 10⁵ Pa.
The balloon is released and rises to a height where its volume becomes 25 m³ and its temperature is 260 K.
Calculate the pressure inside the balloon at its new height. [3 marks]

Ideal revision equation — marks = (pages read × questions answered)²...

All this might sound a bit theoretical, but most gases you'll meet in the everyday world come fairly close to being 'ideal'. They only stop obeying these laws when the pressure's too high or they're getting close to their boiling point.

The Pressure of an Ideal Gas

Kinetic theory tries to **explain** the **gas laws**. *It basically models a gas as a series of hard balls that obey Newton's laws.*

You Can Use Newton's Laws to **Derive** the **Pressure** of an **Ideal Gas**

Imagine a cubic box containing *N* particles of an ideal gas, each with a mass *m*.

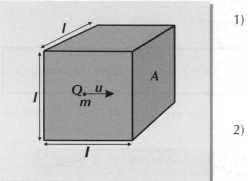

> *See pages 88-89 for a recap of Newton's laws.*

1) The particles of the gas are **free to move** around with **constant random motion**. There are **no forces of attraction** between the particles, so according to **Newton's 1st law**, they continue to move with **constant velocity** until they collide with another particle or the box itself.

2) When a particle **collides** with a **wall** of the box, it exerts a **force** on the wall, and the wall exerts an **equal and opposite force** on the particle. This is **Newton's 3rd law** in action.

3) The **size of the force** exerted by the particle on the wall can be **calculated** using **Newton's 2nd law**, which says the **force is proportional to the rate of change of momentum**.

4) For example, if particle **Q** is travelling directly towards wall **A** with velocity *u*, its momentum is *mu*. When it **hits** the wall, the force of the **impact** causes it to rebound in the **opposite direction**, at the same speed. Its **momentum** is now *–mu*, which means the **change in momentum** is 2*mu*.

5) So, the **force** a particle exerts is **proportional** to its **mass** and its velocity. The **mass** of a single **gas particle** is **tiny** (for example, an atom of helium gas is only 6.6×10^{-27} kg), so each particle can only exert a **minuscule force**.

> Imagine putting a single grain of sugar into a lunchbox and shaking it around — the box might as well be empty. Now put a couple of spoonfuls of sugar in, and shake that around — this time you'll hear the sugar thumping against the walls of the box as it collides. You don't notice the effect of the individual grains, only the combined action of loads of them.

6) But, there **isn't** just **one particle** in the box — there's probably **millions of billions** of them. The **combined force** from so many tiny particles is **much bigger** than the contribution from any individual particle.

7) Because there are so many particles in the box, a **significant number** will be **colliding** with each wall of the box at **any given moment**. And because the particles' motion is **random**, the **collisions** will be **spread** all over the surface of each wall. The result is a **steady**, **even force** on all the walls of the box — this is **pressure**.

Pressure is Proportional to the Number of Particles, their Mass and Speed

So, the pressure in a gas is a result of all the collisions between particles and the walls of the container. How much pressure is exerted by the gas depends on four things:

1) The **volume** of the container — increasing the volume of the container decreases the **frequency of collisions** because the particles have **further to travel** in between collisions. This decreases the pressure.

2) The **number of particles** — increasing the number of particles increases the **frequency of collisions** between the particles and the container, so increases the **total force** exerted by all the collisions.

3) The **mass** of the particles — according to Newton's 2nd law, **force is proportional to mass**, so **heavier** particles will exert a **greater force**.

4) The **speed** of the particles — the **faster** the particles are going when they collide, the **greater** the **change in momentum** and force exerted.

> Or, in symbols: $pV \propto Nm\overline{c^2}$
> ($\overline{c^2}$ is the mean square speed — it represents the square of the speed of a __typical particle__.)

The Pressure of an Ideal Gas

Lots of *Simplifying Assumptions* are Used in *Kinetic Theory*

In **kinetic theory**, physicists picture gas particles moving at **high speed** in **random directions**. To get relations like the one on the previous page, some **simplifying assumptions** are needed:

1) The gas contains a **large number of particles**.
2) The particles **move rapidly** and **randomly**.
3) The motion of the particles follows **Newton's laws**.
4) **Collisions** between particles themselves or between particles and the walls of the container are **perfectly elastic**.
5) There are **no attractive forces** between particles.
6) Any **forces** that act during collisions are **instantaneous**.
7) Particles have a **negligible volume** compared with the volume of the container.

Don't make an ass of yourself — learn the assumptions.

A **gas obeying** these **assumptions** is called an **ideal** gas. Real gases behave like ideal gases as long as the **pressure isn't too big** and the **temperature** is **reasonably high** (compared with their boiling points).

Brownian Motion Supports Kinetic Theory

In 1827, the botanist **Robert Brown** noticed that pollen grains in water constantly moved with a zigzag, **random motion**.

Brownian Motion Experiment

You can **observe** Brownian motion in the lab.

Put some **smoke** in a **brightly illuminated** glass jar and observe the particles using a **microscope**.

The smoke particles appear as **bright specks** moving **haphazardly** from side to side, and up and down.

Brown couldn't explain this, but nearly 80 years later Einstein showed that this provided evidence for the existence of atoms or **molecules** in the air. The **randomly moving** air particles were hitting the smoke particles unevenly, causing this motion.

microscope →

glass cell containing smoke

lamp   glass rod to focus light

Practice Questions

Q1 Describe the connection between collisions and pressure in an ideal gas.
Q2 What are the seven assumptions made about ideal gas behaviour?
Q3 What did Robert Brown observe?
Q4 Why do smoke particles show Brownian motion?

Exam Question

Q1 Brownian motion provides evidence for the continual random motion of particles in a gas.

   (a) Describe an experiment you could use to demonstrate Brownian motion.   [2 marks]

   (b) Explain how Brownian motion supports the theory of the random movement of particles.   [2 marks]

   (c) Outline how the random movement of particles accounts for pressure in a gas.   [1 mark]

Brownian motion — Girl Guide in a tumble-drier...

Pardon me if I'm wrong, but I thought science was all about making predictions and testing them with good old experiments — not having an idea and then creating a load of assumptions to make it work. Hmmm — physicists, eh...

Internal Energy and Temperature

*The energy of a particle depends on its temperature on the **thermodynamic scale** (that's Kelvin to you and me).*

If **A** and **B** are in **Thermal Equilibrium** with **C**, **A** is in **Equilibrium** with **B**

If **body A** and **body B** are both in **thermal equilibrium** with **body C**, then **body A** and **body B** must be in thermal equilibrium with **each other**.

This is linked with the idea of **temperature**.

1) Suppose A, B and C are three identical metal blocks. A has been in a **warm oven**, B has come from a **refrigerator** and C is at **room temperature**.

2) **Thermal energy** flows from A to C and C to B until they all reach **thermal equilibrium** and the net flow of energy stops. This happens when the three blocks are at the **same temperature**.

Thermal energy is **always** transferred from regions of **higher temperature** to regions of **lower temperature**.

The **Speed Distribution** of **Gas Particles** Depends on **Temperature**

The **particles** in a **gas don't** all **travel** at the **same speed**. Some particles will be moving fast but others much more slowly. Most will travel around the average speed. The shape of the **speed distribution** depends on the **temperature** of the gas.

As the temperature of the gas increases:

1) the **average** particle speed increases.

2) the **maximum** particle speed increases.

3) the distribution curve becomes more **spread out**.

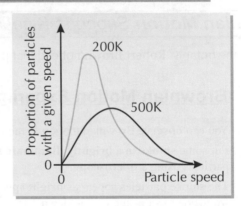

Energy Changes Happen Between Particles

The particles of a gas **collide** with each other **all the time**. Some of these collisions will be '**head-on**' (particles moving in **opposite directions**) while others will be '**shunts from behind**' (particles moving in the **same direction**).

1) As a result of the collisions, **energy** will be **transferred** between particles.

2) Some particles will **gain speed** in a collision and others will **slow down**.

3) **Between collisions**, the particles will travel at **constant speed**.

4) Although the energy of an individual particle changes at each collision, the collisions **don't alter** the **total energy** of the **system**.

5) So, the **average** speed of the particles will stay the same provided the **temperature** of the gas **stays the same**.

Internal Energy is the **Sum** of **Kinetic** and **Potential Energy**

All things (solids, liquids, gases) have **energy** contained within them. The amount of **energy** contained in a system is called its **internal energy** — it's found by **summing** the **kinetic** and **potential energy** of all the **particles** within it.

Internal energy is the **sum** of the **kinetic** and **potential energy** of the **particles** within a system.

For example, the **internal energy** of an **ideal gas** is due to the **kinetic energy** of the **particles** within it. But, how do you **sum** the **individual energies** when the particles all move at **different speeds**, so have **different kinetic energies**? The answer is to find the **average kinetic energy** of a particle (page 107), then **multiply** by the number of particles.

Internal Energy and Temperature

Average Kinetic Energy is Proportional to Absolute Temperature

If you've been paying attention to this section, you'll know that $pV = NkT$ (this is the **equation of state** from page 103) and that pV is **proportional** to $Nm\overline{c^2}$ (page 104). You can **combine** these to find the **kinetic energy** of gas particles.

1) The **equation of state**: $pV = NkT$

2) The **pressure** of an **ideal gas** given by kinetic theory: $pV \propto Nm\overline{c^2}$

   c^2 represents the square of the speed of a typical particle.

3) **Equating** these two gives: $Nm\overline{c^2} \propto NkT$

   N is the number of particles in the gas.

   m is the mass of one particle in the gas.

4) And you can **cancel N** to give: $m\overline{c^2} \propto kT$

5) $\frac{1}{2}m\overline{c^2}$ is the **average kinetic energy** of an **individual particle**. (Remember $\overline{c^2}$ is a measure of speed squared, so this is just like the equation for kinetic energy, $KE = \frac{1}{2}mv^2$, that you'll know already.)

6) This means that kT is **proportional** to **average kinetic energy**:

   The kinetic energy of gas particles is always an average value because the particles are all travelling at different speeds.

   $$E = \frac{3}{2}kT$$

   T is the absolute temperature.

   k is Boltzmann's constant 1.38 × 10⁻²³ JK⁻¹

The **internal energy** of an ideal gas is the **product** of the **average kinetic energy** of its particles and the **number of particles** within it. **Average kinetic energy** is directly proportional to the **absolute temperature** (see above), so **internal energy** must also be **dependent** on **temperature** — a **rise** in the **absolute temperature** will cause an **increase** in the kinetic energy of each particle, meaning a rise in **internal energy**.

Practice Questions

Q1 Describe the changes in the distribution of gas particle speeds as the temperature of a gas increases.

Q2 What is internal energy?

Q3 What would cause a rise in internal energy?

Q4 What happens to the average kinetic energy of a particle if the temperature of a gas doubles?

Exam Questions

Q1 A flask contains one mole of nitrogen molecules at 300 K.

(a) How many molecules of nitrogen are in the flask? [1 mark]

(b) Calculate the average kinetic energy of a nitrogen molecule in the flask. [2 marks]

(c) Explain why all the nitrogen molecules in the flask will not have the same kinetic energy. [1 mark]

Q2 A container of neon gas is cooled until its absolute temperature halves.
Describe and explain any change in the following quantities caused by the cooling:

(i) the average kinetic energy of a molecule of neon gas

(ii) the internal energy of the gas

(iv) the mass of a molecule of neon gas

(iii) the pressure within the container [4 marks]

Positivise your internal energy, man...

*Phew... there's a lot to take in on these pages. Go back over it, step by step, and make sure you understand it all:
the speed distribution of particles, internal energy, average kinetic energy, your A,B,Cs...*

Electric Fields

*Electric fields can be attractive or repulsive, so they're different from gravitational ones. It's all to do with **charge**.*

There is an **Electric Field** around a **Charged Object**

Any object with **charge** has an **electric field** around it — the region where it can attract or repel other charges.

1) Electric charge, **Q**, is measured in **coulombs** (C) and can be either positive or negative.
2) **Oppositely** charged particles **attract** each other. **Like** charges **repel**.
3) If a **charged object** is placed in an electric field, then it will experience a **force**.

You can **Calculate Forces** using **Coulomb's Law**

You'll need **Coulomb's law** to work out **F** — the force of attraction or repulsion between two point charges...

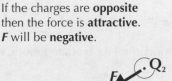

COULOMB'S LAW:

$$F = \frac{kQ_1Q_2}{r^2} \quad \text{where} \quad k = \frac{1}{4\pi\varepsilon}$$

ε ("epsilon") = permittivity of material between charges
Q_1 and Q_2 are the charges
r is the distance between Q_1 and Q_2

If the charges are **opposite** then the force is **attractive**. **F** will be **negative**.

If Q_1 and Q_2 are **like** charges then the force is **repulsive**, and **F** will be **positive**.

1) The force on Q_1 is always **equal** and **opposite** to the force on Q_2.
2) It's an **inverse square law**. Again. The further apart the charges are, the weaker the force between them.
3) The size of the force **F** also depends on the **permittivity**, ε, of the material between the two charges. For free space, the permittivity is $\varepsilon_0 = 8.85 \times 10^{-12}\,\text{C}^2\text{N}^{-1}\text{m}^{-2}$.

Electric Field Strength is Force per Unit Charge

Electric field strength, **E**, is defined as the **force per unit positive charge** — the force that a charge of +1 C would experience if it was placed in the electric field.

$$E = \frac{F}{q}$$

F is the force on a 'test' charge **q**.

1) **E** is a **vector** pointing in the **direction** that a **positive charge** would **move**.
2) The units of **E** are **newtons per coulomb** (NC^{-1}).
3) Field strength depends on **where you are** in the field.
4) A **point charge** — or any body which behaves as if all its charge is concentrated at the centre — has a **radial** field.

In a **Radial Field**, E is **Inversely Proportional** to r^2

1) **E** is the force per unit charge that a small, positive 'test' charge, **q**, would feel at different points in the field. In a **radial field**, **E** depends on the distance **r** from the point charge **Q**...

$$E = \frac{kQ}{r^2} \quad \left(k = \frac{1}{4\pi\varepsilon}\right)$$

For a **positive Q**, the small positive 'test' charge **q** would be **repelled**, so the field lines point **away** from **Q**.

For a **negative Q**, the small positive charge **q** would be **attracted**, so the field lines point **towards Q**.

2) It's another **inverse square law** — $E \propto \dfrac{1}{r^2}$

3) Field strength **decreases** as you go **further away** from **Q** — on a diagram, the **field lines** get **further apart**.

The area under the graph is the electric potential.

Electric Fields

Field Strength is the Same Everywhere in a Uniform Field

A **uniform field** can be produced by connecting two **parallel plates** to the opposite poles of a battery.

1) Field strength **E** is the **same** at **all points** between the two plates and is...

$$E = \frac{V}{d}$$

V is the **potential difference** between the plates
d is the distance between them

2) **E** can be measured in volts per metre (Vm^{-1}).

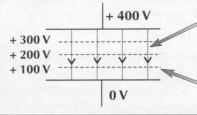

The **lines of force** are **parallel** to each other.

Areas with the **same potential** are **parallel** to the **plates**, and **perpendicular** to the **field lines**.

Charged Particles Move Through Uniform Electric Fields Like Projectiles

You'll probably remember from AS that **projectiles** move through a **uniform gravitational field** along a curved path called a **parabola**. **Charged particles** do a similar thing when they move through **uniform electric fields**.

1) A charged particle will experience a **constant force parallel** to the **electric field lines**.

2) If the particle is **positively charged** then the force is in the **same direction** as the field lines.
If it's **negatively charged** (e.g. an **electron**), the force is in the **opposite direction** to the field lines.

3) The particle will **accelerate at a constant rate** in the direction of this force — that's just **Newton's second law**.

4) If the particle's **velocity** has a **component** at **right angles** to the field lines, the particle will keep moving in this direction with a **uniform velocity**. That's **Newton's first law**.

5) The combined effect of constant acceleration in one direction and constant velocity at right angles is a **parabola**.

There are Similarities between Gravitational and Electric Fields...

1)	Gravitational field strength, **g**, is **force** per **unit mass**.	Electric field strength, **E**, is **force** per **unit positive charge**.
2)	Newton's law of gravitation for the **force** between two point masses is an **inverse square law**. $F \propto \frac{1}{r^2}$	Coulomb's law for the electric **force** between two point charges is also an **inverse square law**. $F \propto \frac{1}{r^2}$
3)	A point mass creates a **radial gravitational field**.	A point charge creates a **radial electric field**.

... and some Differences too

1) Gravitational forces are always **attractive**. Electric forces can be either **attractive** or **repulsive**.

2) Objects can be **shielded** from **electric** fields, but not from gravitational fields.

3) The **medium** between the charges or masses affects the **size** of **electric forces**, but not gravitational forces.

Practice Questions

Q1 Draw the electric field lines due to a positive charge, and due to a negative charge.
Q2 Write down Coulomb's law.

Exam Questions

Q1 The diagram shows two electric charges with equal but opposite charge. $Q = 1.6 \times 10^{-19}$ C.
(a) Draw the electric field lines to show the electric field in the area surrounding the charges. [3 marks]
(b) The distance between the charges is 3.5×10^{-10} m. Find the strength of the electric field halfway between the charges. [3 marks]

3.5×10^{-10} m
+Q   -Q

Q2 (a) Two parallel plates are separated by an air gap of 4.5 mm. The plates are connected to a 1500 V dc supply. What is the electric field strength between the plates? Give a suitable unit and state the direction of the field. [3 marks]
(b) The plates are now pulled further apart so that the distance between them is doubled. The electric field strength remains the same. What is the new voltage between the plates? [2 marks]

Electric fields — one way to roast beef...

At least you get a choice here — uniform or radial, positive or negative, attractive or repulsive, chocolate or strawberry...

Magnetic Fields

Magnetic fields — making pretty patterns with iron filings before spending an age trying to pick them off the magnet.

A **Magnetic Field** is a **Region** Where a **Force** is Exerted on **Magnetic Materials**

1) Magnetic fields can be represented by **field lines**.
2) Field lines go from **north to south**.
3) The **closer** together the lines, the **stronger** the field.
4) At a **neutral point**, the magnetic fields **cancel out**.

A neutral point

There is a **Magnetic Field** Around a **Wire** Carrying **Electric Current**

1) When **current flows** in a wire (or any other long straight conductor), there's a **magnetic field** around the wire.

2) The **field lines** form **concentric circles** centred on the wire — the **closer** together the circles are, the **stronger** the field.

3) The **direction** of a magnetic **field** around a current-carrying wire can be worked out with the **right-hand rule**.

4) If you wrap the wire around to make a **single coil** or a **solenoid**, the shape of the **field changes** as the lines interact.

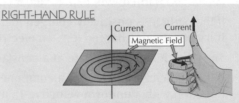

RIGHT-HAND RULE

1) Stick your right thumb up, like you're hitching a lift.
2) If your thumb points in the direction of the current...
3) ...your curled fingers point in the direction of the field.

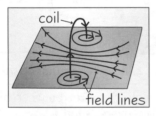

coil

field lines

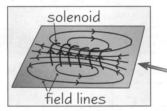

solenoid

field lines

The magnetic field around a solenoid is the same shape as the one around a bar magnet.

A **Wire** Carrying a **Current** in a **Magnetic Field** will **Experience** a **Force**

1) If you put a **current-carrying wire** into an **external** magnetic field (e.g. between two magnets), the field around the wire and the field from the magnets **interact**. The field lines from the magnet **contract** to form a **'stretched catapult'** effect where the flux lines are closer together.

2) This causes a **force** on the wire.

3) If the current is **parallel** to the flux lines, **no force** acts.

4) The **direction** of the force is always **perpendicular** to both the **current** direction and the **magnetic field** — it's given by **Fleming's left-hand rule**...

Resulting Force

N   S

→ Normal magnetic field of wire
→ Normal magnetic field of magnets
→ Deviated magnetic field of magnets

The **Direction** of the **Force** is Given by **Fleming's Left-Hand Rule**

You can work out the direction of the **force** on a **current-carrying wire** using **Fleming's left-hand rule** — just remember to use your **left hand**...

Fleming's Left-Hand Rule

The **F**irst finger points in the direction of the uniform magnetic **F**ield, the se**C**ond finger points in the direction of the conventional **C**urrent. Then your thu**M**b points in the direction of the force (in which **M**otion takes place).

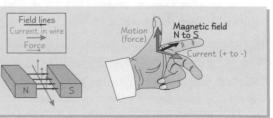

Field lines
Current in wire
Force

Motion (force)

Magnetic field N to S

Current (+ to -)

Magnetic Fields

The **Size** of the **Force** can be **Calculated** using **F = BIl**...

1) The size of the **force**, **F**, on a current-carrying wire at right-angles to a magnetic field is proportional to the **current**, **I**, the **length of wire** in the field, **l**, and the **strength of the magnetic field**, **B**. This gives the equation:

$$F = BIl$$

2) In this equation, the **magnetic field strength**, **B**, is defined as:

> The **force** on **one metre** of wire carrying a **current** of **one amp** at **right angles** to the **magnetic field**.

3) **Magnetic field strength** is also called **flux density** and it's measured in **teslas**, **T**.

$$1 \text{ tesla} = \frac{\text{Wb}}{\text{m}^2}$$

It helps to think of flux density as the number of flux lines (measured in webers (Wb), see p 112) per unit area.

4) Magnetic field strength is a **vector** quantity with both a **direction** and **magnitude**.

The Force is **Greatest** when the **Wire** and **Field** are **Perpendicular**...

1) The **force** on a current-carrying wire in a magnetic field is caused by the **component** of field strength which is **perpendicular** to the wire — **B** sin **θ**.

2) So, for a wire at an **angle θ** to the field, the **force** acting on the wire is given by:

$$F = BIl \sin \theta$$

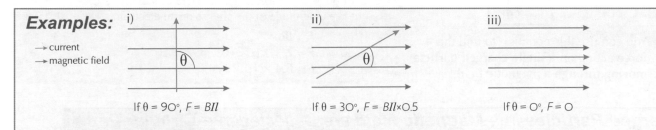

Examples:

i) If θ = 90°, F = BIl

ii) If θ = 30°, F = BIl×0.5

iii) If θ = 0°, F = 0

→ current
→ magnetic field

Practice Questions

Q1 Describe why a current-carrying wire at right angles to an external magnetic field will experience a force.

Q2 Write down the equation you would use to find the force on a current-carrying wire that is at an angle of 30° to an external uniform field.

Q3 Sketch the magnetic fields around a long straight current-carrying wire and a solenoid. Show the direction of the current and magnetic field on each diagram.

Q4 A copper bar can roll freely on two copper supports, as shown in the diagram. When current is applied in the direction shown, which way will the bar roll?

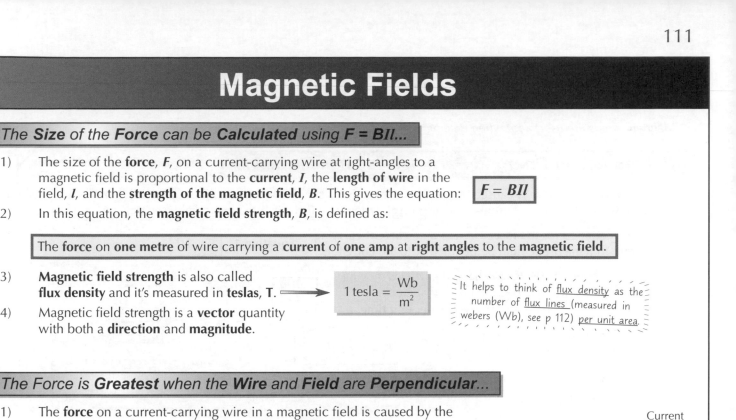

Horseshoe magnet

Copper bar

Exam Question

Q1 A 4 cm length of wire carrying a current of 3 A runs perpendicular to a magnetic field of strength 2×10^{-5} T.

(a) Calculate the magnitude of the force on the wire. [2 marks]

(b) If the wire is rotated so that it is at 30° to the direction of the field, what would the size of the force be? [2 marks]

I revised the right hand rule by the A69 and ended up in Newcastle...

Fleming's left-hand rule is the key to this section — so make sure you know how to use it and understand what it all means. Remember that the direction of the magnetic field is from N to S, and that the current is from +ve to –ve — this is as important as using the correct hand. You need to get those right or it'll all go to pot...

Charged Particles in Magnetic Fields

Magnetic fields are used a lot when dealing with particle beams.

Forces Act on Charged Particles in Magnetic Fields

Electric current in a wire is caused by the **flow** of negatively **charged** electrons. These charged particles are affected by **magnetic fields** — so a current-carrying wire experiences a **force** in a magnetic field (see pages 110–111).

1) The equation for the **force** exerted on a **current-carrying wire** in a **magnetic field** perpendicular to the current is:

Equation 1: $F = BIl$

2) To see how this relates to **charged particles** moving through a wire, you need to know that electric **current**, I, is the flow of **charge**, q, per unit **time**, t:

$$I = \frac{q}{t}$$

3) A charged particle which moves a **distance** l in **time** t has a **velocity**, v:

$$v = \frac{l}{t} \Rightarrow t = \frac{l}{v}$$

In many exam questions, q is the size of the charge on the electron, which is 1.6×10^{-19} coulombs.

4) So, putting the two equations **together** gives the **current** in terms of the **charge** flowing through the **wire**:

Equation 2: $I = \frac{qv}{l}$

5) Putting **equation 2** back into **equation 1** gives the **electromagnetic force** on the wire as:

$$F = Bqv$$

6) You can use this equation to find the **force** acting on a **single charged particle** moving through a magnetic field.

Example

What is the force acting on an electron travelling at 2×10^4 ms^{-1} through a uniform magnetic field of strength 2 T? (The magnitude of the charge on an electron is 1.6×10^{-19} C.)

$F = Bqv$

so, $F = 2 \times 1.6 \times 10^{-19} \times 2 \times 10^4$

so, $F = 6.4 \times 10^{-15}$ N

Charged Particles in a Magnetic Field are Deflected in a Circular Path

1) By **Fleming's left-hand rule** the force on a **moving charge** in a magnetic field is always **perpendicular** to its **direction of travel**.

2) Mathematically, that is the condition for **circular** motion.

3) This effect is used in **particle accelerators** such as **cyclotrons** and **synchrotrons**, which use **magnetic fields** to accelerate particles to very **high energies** along circular paths.

4) The **radius of curvature** of the **path** of a charged particle moving through a magnetic field gives you information about the particle's **charge** and **mass** — this means you can **identify different particles** by studying how they're **deflected** (see next page).

Circular path

Force on particle

Positively charged particle

Magnetic field into paper (⊗)

Centripetal Force Tells Us About a Particle's Path

The centripetal force and the electromagnetic force are equivalent for a charged particle travelling along a circular path.

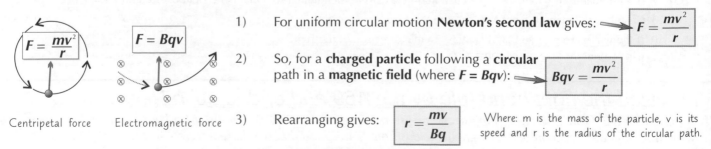

$F = \frac{mv^2}{r}$

$F = Bqv$

Centripetal force   Electromagnetic force

1) For uniform circular motion **Newton's second law** gives: $F = \frac{mv^2}{r}$

2) So, for a **charged particle** following a **circular** path in a **magnetic field** (where $F = Bqv$): $Bqv = \frac{mv^2}{r}$

3) Rearranging gives: $r = \frac{mv}{Bq}$

Where: m is the mass of the particle, v is its speed and r is the radius of the circular path.

Charged Particles in Magnetic Fields

Scientists use Mass Spectrometers to Analyse Samples

1) **Mass spectrometers** are used to **analyse samples** to find out what **chemicals** are present within them and the **relative proportions** of each one.

2) First, the sample is **vaporised** (i.e. turned to gas) and **ionised** before passing along the **spectrometer tube**.

3) The ions travel through **electric** and **magnetic fields** to a **detector**, which is connected to a **computer**.

4) The **identity** of the **ions** reaching the **detector** can be determined from their **mass to charge ratio**.

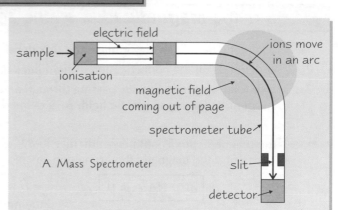

A Mass Spectrometer

The Electric Field Accelerates the Ions...

1) The **ions** are **charged**, so they experience a **force** as they travel through the **electric field**.

2) This force **accelerates** the ions along the **evacuated spectrometer tube** towards the magnetic field.

3) The **velocity** of the ions depends on the **electric field strength** — **increasing** the **field strength** increases the **force** on the ions, so **increases** their **acceleration** and final **velocity**.

... And the Magnetic Field Deflects the Ions in a Circular Path

1) The **magnetic field** deflects the ions in a **circular path** (see the previous page).

2) The **radius** of the path of each **ion** depends on its **mass to charge ratio**, its **velocity** and the strength of the **magnetic field** — see the equation at the bottom of the previous page.

3) The **velocity** of the ions and the **magnetic field strength** can be **controlled** by the user.

4) Only the **ions** moving in an arc of a particular **radius** can pass through the slit into the **detector**.

5) This means that only ions with a particular **mass to charge ratio** reach the detector at any one time.

6) The detector is connected to a **computer**, which **identifies the ions** from their **mass to charge ratio** and the spectrometer settings, and records the **amount** of each ion.

The field had deflected the lions onto a circular path.

Practice Questions

Q1 Derive the formula for the force on a charged particle in a magnetic field, $F = Bqv$, from $F = BIl$.

Q2 Give two examples of how magnetic fields can be used.

Q3 Outline the role of the electric and magnetic fields in a mass spectrometer.

Exam Questions

Q1 (a) What is the force on an electron travelling at a velocity of 5×10^6 ms^{-1} through a perpendicular magnetic field of 0.77 T? [The charge on an electron is -1.6×10^{-19} C.] [2 marks]

  (b) Explain why it follows a circular path while in the field. [1 mark]

Q2 What is the radius of the circular path of an electron with a velocity of 2.3×10^7 ms^{-1} moving perpendicular to a magnetic field of 0.6 mT?
[The mass of an electron is 9.11×10^{-31} kg and its charge is -1.6×10^{-19} C.] [3 marks]

Q3 A sample of sodium chloride is analysed using a mass spectrometer. The magnetic field is initially set to 0.20 T and ions of the isotope Cl-35 (mass 35 u) reach the detector. What magnetic field strength would you need for Cl-37 ions (mass 37 u) to reach the detector? Assume that the detector remains in the same place, the electric field remains constant and that both types of ion have the same size charge as an electron. [3 marks]

Hold on to your hats folks — this is starting to get tricky...

Basically, the main thing you need to know here is that both electric and magnetic fields will exert a force on a charged particle. There's even a handy equation to work out the force on a charged particle moving through a magnetic field — it might not impress your friends, but it will impress the examiner, so learn it.

Electromagnetic Induction

Producing electricity by waggling a wire about in a magnetic field sounds like monkey magic — but it's real physics...

Think of the **Magnetic Flux** as the Total **Number** of **Field Lines**...

1) Magnetic field strength, or **magnetic flux density**, **B**, is a measure of the **strength** of a magnetic field **per unit area**.

2) It depends on the concentration of **field lines** or **magnetic flux** passing through that area.

3) The total **magnetic flux**, ϕ, passing through an **area**, **A**, perpendicular to a **magnetic field**, **B**, is defined as: $\longrightarrow$ $\boxed{\phi = BA}$

The unit of ϕ ('phi') is the weber — see below for a definition.

If the magnetic flux is **not perpendicular** to **B**, you can find the **magnetic flux** using this equation:

$$\phi = BA \cos \theta$$

where θ is the **angle** between the **field** and the **normal** to the plane of the coil.

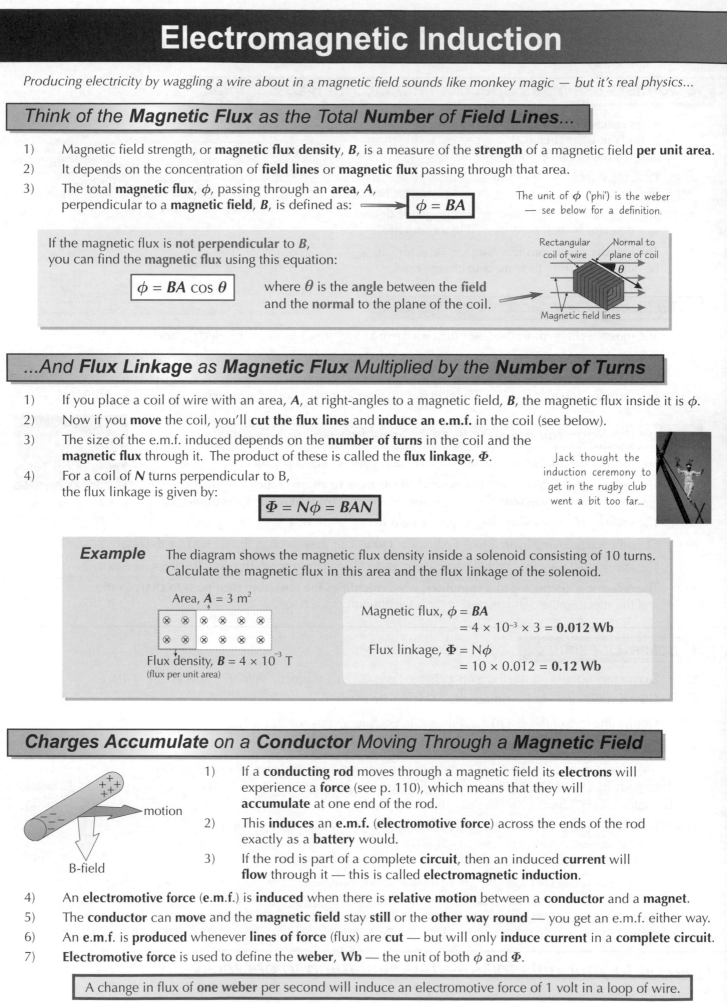

Rectangular coil of wire   Normal to plane of coil θ

Magnetic field lines

...And **Flux Linkage** as **Magnetic Flux** Multiplied by the **Number of Turns**

1) If you place a coil of wire with an area, **A**, at right-angles to a magnetic field, **B**, the magnetic flux inside it is ϕ.

2) Now if you **move** the coil, you'll **cut the flux lines** and **induce an e.m.f.** in the coil (see below).

3) The size of the e.m.f. induced depends on the **number of turns** in the coil and the **magnetic flux** through it. The product of these is called the **flux linkage**, Φ.

4) For a coil of **N** turns perpendicular to B, the flux linkage is given by:

$$\Phi = N\phi = BAN$$

Jack thought the induction ceremony to get in the rugby club went a bit too far...

Example   The diagram shows the magnetic flux density inside a solenoid consisting of 10 turns. Calculate the magnetic flux in this area and the flux linkage of the solenoid.

Area, **A** = 3 m^2

⊗ ⊗ ⊗ ⊗ ⊗ ⊗
⊗ ⊗ ⊗ ⊗ ⊗ ⊗

Flux density, **B** = 4 × 10^{-3} T
(flux per unit area)

Magnetic flux, $\phi = BA$
     = 4 × 10^{-3} × 3 = **0.012 Wb**

Flux linkage, $\Phi = N\phi$
     = 10 × 0.012 = **0.12 Wb**

Charges Accumulate on a **Conductor** Moving Through a **Magnetic Field**

+ + +
+ + +
— — —   motion
— —

B-field

1) If a **conducting rod** moves through a magnetic field its **electrons** will experience a **force** (see p. 110), which means that they will **accumulate** at one end of the rod.

2) This **induces** an **e.m.f. (electromotive force)** across the ends of the rod exactly as a **battery** would.

3) If the rod is part of a complete **circuit**, then an induced **current** will **flow** through it — this is called **electromagnetic induction**.

4) An **electromotive force** (**e.m.f.**) is **induced** when there is **relative motion** between a **conductor** and a **magnet**.

5) The **conductor** can **move** and the **magnetic field** stay **still** or the **other way round** — you get an e.m.f. either way.

6) An **e.m.f.** is **produced** whenever **lines of force** (flux) are **cut** — but will only **induce current** in a **complete circuit**.

7) **Electromotive force** is used to define the **weber**, **Wb** — the unit of both ϕ and Φ.

> A change in flux of **one weber** per second will induce an electromotive force of 1 volt in a loop of wire.

Electromagnetic Induction

These Results are Summed up by Faraday's Law...

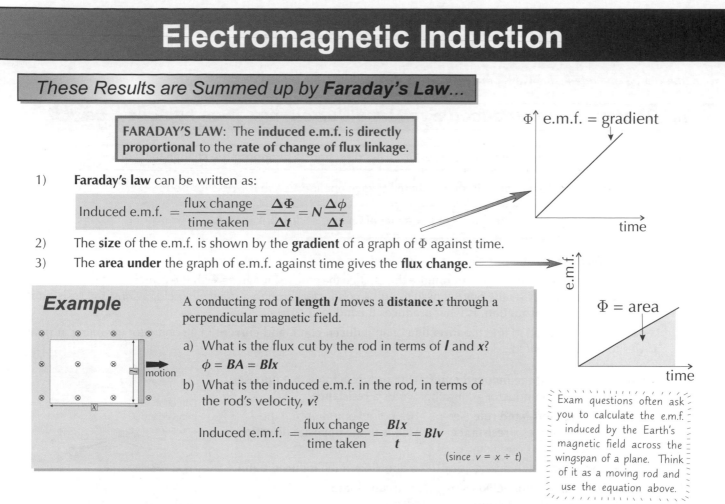

FARADAY'S LAW: The **induced e.m.f.** is **directly proportional** to the **rate of change of flux linkage**.

1) **Faraday's law** can be written as:

$$\text{Induced e.m.f.} = \frac{\text{flux change}}{\text{time taken}} = \frac{\Delta\Phi}{\Delta t} = N\frac{\Delta\phi}{\Delta t}$$

2) The **size** of the e.m.f. is shown by the **gradient** of a graph of Φ against time.

3) The **area under** the graph of e.m.f. against time gives the **flux change**.

e.m.f. = gradient

Φ = area

Example

A conducting rod of **length l** moves a **distance x** through a perpendicular magnetic field.

a) What is the flux cut by the rod in terms of l and x?

$$\phi = BA = Blx$$

b) What is the induced e.m.f. in the rod, in terms of the rod's velocity, v?

$$\text{Induced e.m.f.} = \frac{\text{flux change}}{\text{time taken}} = \frac{Blx}{t} = Blv$$

(since $v = x \div t$)

motion

Exam questions often ask you to calculate the e.m.f. induced by the Earth's magnetic field across the wingspan of a plane. Think of it as a moving rod and use the equation above.

Practice Questions

Q1 What is the difference between magnetic flux density, magnetic flux and magnetic flux linkage?

Q2 State Faraday's law.

Q3 A coil consists of N turns each of area A. If it is placed at right angles to a uniform magnetic field, what is its flux linkage?

Q4 Explain how you can find the direction of an induced e.m.f. in a copper bar moving at right angles to a magnetic field.

Exam Questions

Q1 A coil of area 0.23 m² is placed at right angles to a magnetic field of 2×10^{-3} T.

(a) What is the magnetic flux passing through the coil? [2 marks]

(b) If the coil has 150 turns, what is the magnetic flux linkage in the coil? [2 marks]

(c) Over a period of 2.5 seconds the magnetic field is reduced uniformly to 1.5×10^{-3} T. What is the size of the e.m.f. induced across the ends of the coil? [3 marks]

Q2 A 0.01 m² coil of 500 turns is perpendicular to a magnetic field of 0.9 T.

(a) What is the magnetic flux linkage in the coil? [2 marks]

(b) The coil is rotated until the normal to the plane of the coil is at 90° to the direction of the magnetic field. The movement is uniform and takes 0.5 s. Calculate the e.m.f. induced by this movement. [4 marks]

Beware — physics can induce extreme confusion...

OK... I know that might have seemed a bit scary... but the more you go through it, the more it stops being a big scary monster of doom and just becomes another couple of equations you have to remember. Plus it's one of those things that makes you sound well clever... "What did you learn today, Jim?", "Oh, just magnetic flux linkage in solenoids, Mum..."

Electromagnetic Induction

E.m.f. is a bit of a rebel... it always opposes the change that caused it.

The **Direction** of the **Induced E.m.f.** and **Current** are given by **Lenz's Law**

> **LENZ'S LAW:** The **induced e.m.f.** is always in such a **direction** as to **oppose** the **change** that caused it.

1) **Lenz's law** and **Faraday's law** can be **combined** to give one formula that works for both:

> induced e.m.f. = − rate of change of flux linkage

2) The **minus sign** shows the direction of the **induced e.m.f.**

3) The idea that an induced e.m.f. will **oppose** the change that caused it agrees with the principle of the **conservation of energy** — the **energy used** to pull a conductor through a magnetic field, against the **resistance** caused by magnetic **attraction**, is what **produces** the **induced current**.

4) **Lenz's law** can be used to find the **direction** of an **induced e.m.f.** and **current** in a conductor travelling at right angles to a magnetic field...

1) **Lenz's law** says that the **induced e.m.f.** will produce a force that **opposes** the motion of the conductor — in other words a **resistance**.

2) Using **Fleming's left-hand rule** (see p. 110), point your thumb in the direction of the force of **resistance** — which is in the **opposite direction** to the motion of the conductor.

3) Your **second finger** will now give you the direction of the **induced e.m.f.**

4) If the conductor is **connected** as part of a **circuit**, a current will be induced in the **same direction** as the induced e.m.f.

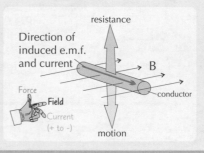

An **Alternator** is a **Generator** of **Alternating Current**

1) **Generators**, or dynamos, **convert** kinetic energy into **electrical energy** — they **induce** an electric **current** by **rotating** a **coil** in a magnetic field.

2) The diagram shows a simple **alternator** — a generator of **AC**. It has **slip rings** and **brushes** to connect the coil to an external circuit.

3) The output **voltage** and **current** change direction with every **half rotation** of the coil, producing **alternating current** (**AC**).

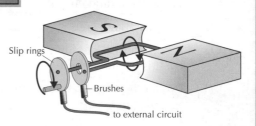

Transformers Work by Electromagnetic **Induction**

1) **Transformers** are devices that use electromagnetic induction to **change** the size of the **voltage** for an **alternating current**.

2) They consist of **two coils of wire** wrapped around an **iron core**.

3) An alternating current flowing in the **primary** (or input) **coil** produces **magnetic flux** in the **iron core**.

4) The **magnetic field** is passed through the **iron core** to the **secondary** (or output) coil, where it **induces** an alternating **voltage** of the same frequency.

5) The **ratio** of the **number of turns** on each coil along with the voltage across the primary coil determines the **size of the voltage** induced in the secondary coil.

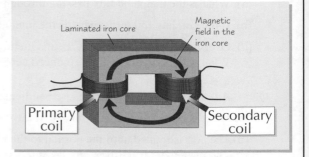

> **Step-up** transformers **increase** the **voltage** by having **more turns** on the **secondary** coil than the primary.
>
> **Step-down** transformers **reduce** the voltage by having **fewer** turns on the secondary coil.

Electromagnetic Induction

You Can *Calculate* the *Voltage Change* Induced by a *Transformer*

From Faraday's law (page 115), the **induced** e.m.f.s in both the **primary** and **secondary** coils can be calculated:

Primary coil
$$V_p = N_p \frac{\Delta \Phi}{\Delta t}$$

Secondary coil
$$V_s = N_s \frac{\Delta \Phi}{\Delta t}$$

⟶ These can be combined to give the equation for an **ideal transformer**: ⟶

$$\frac{V_p}{V_s} = \frac{N_p}{N_s}$$

(where N is the number of turns in a coil)

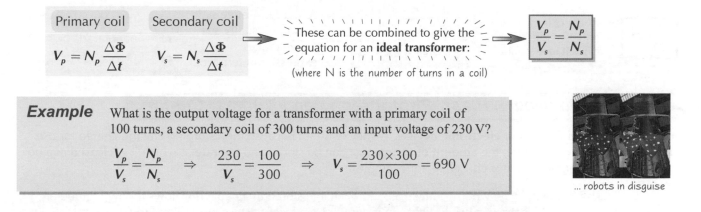

Example   What is the output voltage for a transformer with a primary coil of 100 turns, a secondary coil of 300 turns and an input voltage of 230 V?

$$\frac{V_p}{V_s} = \frac{N_p}{N_s} \quad \Rightarrow \quad \frac{230}{V_s} = \frac{100}{300} \quad \Rightarrow \quad V_s = \frac{230 \times 300}{100} = 690 \text{ V}$$

... robots in disguise

Transformers are an *Important* Part of the *National Grid*

1) **Electricity** from power stations is sent round the country in the **national grid** at the **lowest** possible current, because **losses** due to the **resistance** in the cables are proportional to the current squared.

2) Since **power = current × voltage**, a **low current** means a **high voltage**.

3) **Transformers** allow us to **step up** the voltage to around **400 000 V** for **transmission** through the national grid, and then **reduce** it again to **230 V** for domestic use.

Practice Questions

Q1   State Lenz's law.

Q2   Explain how you can find the direction of an induced e.m.f. in a copper bar moving at right angles to a magnetic field.

Q3   What is the function of an AC generator? Outline how one works.

Q4   Draw a diagram of a simple transformer. What is meant by a step-down transformer?

Exam Questions

Q1   An aeroplane with a wingspan of 30 m flies at a speed of 100 ms⁻¹ perpendicular to the Earth's magnetic field as shown. The Earth's magnetic field at the aeroplane's location is 60×10^{-6} T.

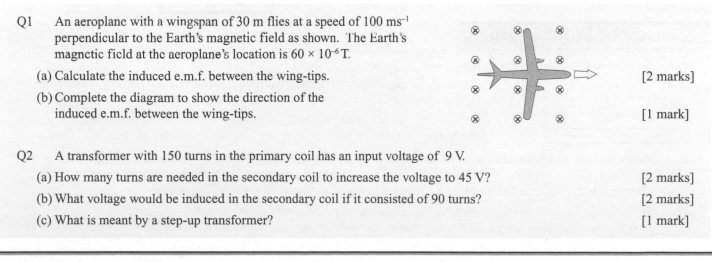

(a) Calculate the induced e.m.f. between the wing-tips.   [2 marks]

(b) Complete the diagram to show the direction of the induced e.m.f. between the wing-tips.   [1 mark]

Q2   A transformer with 150 turns in the primary coil has an input voltage of 9 V.

(a) How many turns are needed in the secondary coil to increase the voltage to 45 V?   [2 marks]

(b) What voltage would be induced in the secondary coil if it consisted of 90 turns?   [2 marks]

(c) What is meant by a step-up transformer?   [1 mark]

Aaaaaand — relax...

Breathe a sigh of relief, pat yourself on the back and make a brew — well done, you've reached the end of the section. That was pretty nasty stuff, but don't let all of those equations get you down — once you've learnt the main ones and can use them blindfolded, even the trickiest looking exam question will be a walk in the park...

Capacitors

Capacitors are things that store electrical charge — like a charge bucket. Sounds simple enough... ha... ha, ha, ha...

Capacitance is Defined as the Amount of Charge Stored per Volt

$$C = \frac{Q}{V}$$

where **Q** is the **charge** in coulombs, **V** is the **potential difference** in volts and **C** is the **capacitance** in farads (F) — 1 farad = 1 C V⁻¹.

A farad is a **huge** unit so you'll usually see capacitances expressed in terms of:

μF — microfarads (× 10^{-6})

nF — nanofarads (× 10^{-9})

pF — picofarads (× 10^{-12})

Capacitors are used in Flash Photography and Nuclear Fusion

Capacitors are found in loads of **electronic devices** — they're useful because they **store** up electric charge for use when you want it. What's more, the **amount of charge** that can be stored and the **rate** at which it's **released** can be controlled by the type of capacitor chosen — for example:

1) **Flash photography** — when you take a picture, charge stored in a **capacitor** flows through a tube of xenon gas which emits a **bright light**. In order to give a **brief flash** of bright light, the capacitor has to discharge really quickly to give a **short pulse** of high current.

2) **Nuclear fusion** — a colossal amount of **energy** is needed to start a nuclear fusion reaction (p. 143) — one way to deliver this is using **lasers** controlled by **capacitors**. The capacitors release a huge charge in a fraction of a second to **maximise** the amount of energy in each pulse of laser-light.

3) **Back-up power supplies** — **computers** are often connected to back-up power supplies to make sure that you don't lose any data if there's a **power cut**. These often use **large capacitors** that store charge while the power is on then release that charge slowly if the power goes off. The capacitors are designed to discharge over a number of hours, maintaining a **steady flow** of charge.

Combined Capacitance Increases in Parallel, but Decreases in Series

1) If you put two or more **capacitors** in a **parallel circuit**, the **p.d.** across each one is the **same**.

2) Each capacitor can store the **same** amount of **charge** as if it was the **only component** in the circuit.

3) So, the **total capacitance** is just the **sum** of the individual capacitances:

$$C_{total} = C_1 + C_2$$

4) When you put capacitors in a **series circuit**, the **p.d.** is **shared** between each one.

5) Each capacitor can only store a **fraction** of the **charge** available, so the **total capacitance** is **less than** the individual capacitances:

$$\frac{1}{C_{total}} = \frac{1}{C_1} + \frac{1}{C_2}$$

You could be asked to solve circuit problems using these equations.

Capacitors Store Energy

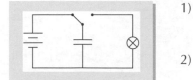

1) In this circuit, when the switch is flicked to the **left**, **charge** builds up on the plates of the **capacitor**. **Electrical energy**, provided by the battery, is **stored** by the capacitor.

2) If the switch is flicked to the **right**, the energy stored on the plates will **discharge** through the **bulb**, converting electrical energy into light and heat.

3) **Work** is done **removing charge** from **one plate** and depositing **charge** onto the other one to charge the capacitor. The energy for this must come from the **electrical energy** of the **battery**, and is given by **charge × p.d.** The energy **stored** by a capacitor is **equal** to the **work done** by the **battery**.

4) So, you can find the **energy stored** by the capacitor from the **area** under a **graph** of **p.d.** against **charge stored** on the capacitor. The p.d. across the capacitor is **proportional** to the charge stored on it, so the graph will be a **straight line** through the origin. The **energy stored** is given by the **yellow triangle**.

5) **Area of triangle = ½ × base × height**, so the energy stored by the capacitor is:

$$W = \frac{1}{2}QV$$

You need to remember where this equation comes from.

W stands for 'work done', but you can also use **E** for energy

Capacitors

There are **Three** Expressions for the **Energy Stored** by a Capacitor

1) You know the first one already: $\boxed{W = \dfrac{1}{2}QV}$ ← **W** stands for 'work done', but you can also use **E** for energy

2) $C = \dfrac{Q}{V}$, so $Q = CV$. Substitute that into the energy equation: $W = \dfrac{1}{2}CV \times V$. So: $\boxed{W = \dfrac{1}{2}CV^2}$

3) $V = \dfrac{Q}{C}$, so $W = \dfrac{1}{2}Q \times \dfrac{Q}{C}$. Simplify: $\boxed{W = \dfrac{Q^2}{2C}}$

Example

A 900 µF capacitor is charged up to a potential difference of 240 V. Calculate the energy stored by the capacitor.

First, choose the best equation to use — you've been given **V** and **C**, so you need $W = \dfrac{1}{2}CV^2$.

Substitute the values in: $E = \dfrac{1}{2} \times 9 \times 10^{-4} \times 240^2 = 25.92$ J

Practice Questions

Q1 Define capacitance.

Q2 What is the relationship between charge, voltage and capacitance?

Q3 Write the following in standard form: a) 220 µF b) 1000 pF c) 470 nF.

Exam Questions

Q1 Capacitors are found in many electronic devices, including cameras.

(a) Outline how capacitors are used in cameras. [1 mark]

(b) Explain why a capacitor is a suitable component for this use. [2 marks]

Q2 From the graphs below, calculate the capacitance of the capacitor and the charge stored on its plates.

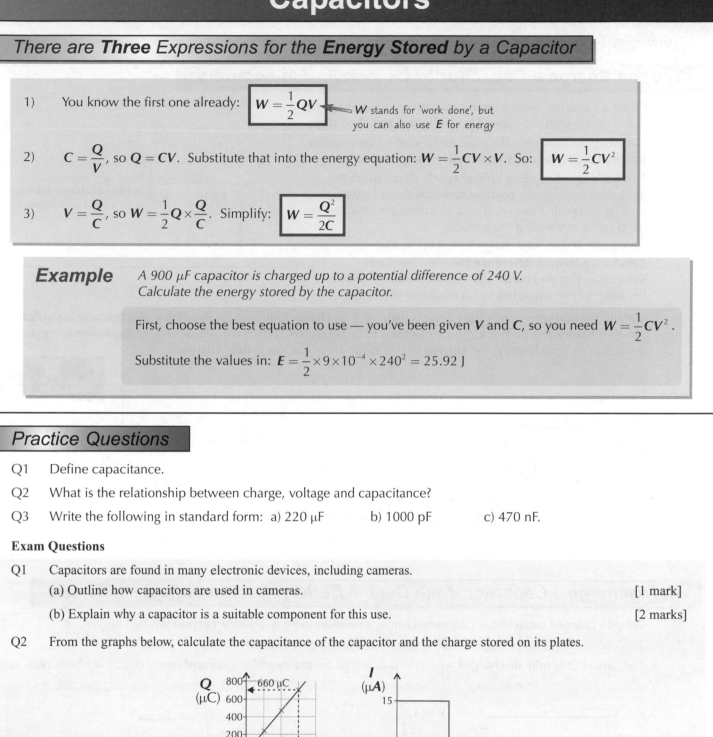

[4 marks]

Q3 A 500 mF capacitor is fully charged up from a 12 V supply.

(a) Calculate the total energy stored by the capacitor. [2 marks]

(b) Calculate the charge stored by the capacitor. [2 marks]

Q4 Explain why a circuit with two 470 mF capacitors in parallel would store more charge than a circuit with two 470 mF capacitors in series, when the same voltage is applied to both circuits.
In your answer, you should make clear the sequence of steps in your argument. [4 marks]

Capacitance — fun, it's not...

Capacitors are really useful in the real world. Pick an appliance, any appliance, and it'll probably have a capacitor or several. If I'm being honest though, the only saving grace of these pages for me is that they're not especially hard...

Charging and Discharging

Charging and discharging — sounds painful...

You can **Charge** a **Capacitor** by **Connecting it** to a **Battery**

1) When a capacitor is connected to a **battery**, a **current** flows in the circuit until the capacitor is **fully charged**, then **stops**.

2) The electrons flow onto the plate connected to the **negative terminal** of the battery, so a **negative charge** builds up.

3) This build-up of negative charge **repels** electrons off the plate connected to the **positive terminal** of the battery, making that plate positive. These electrons are attracted to the positive terminal of the battery.

4) An **equal** but **opposite** charge builds up on each plate, causing a **potential difference** between the plates. Remember that **no charge** can flow **between** the plates because they're **separated** by an **insulator** (dielectric).

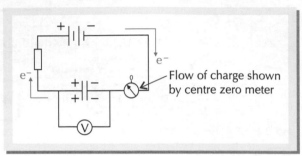

Flow of charge shown by centre zero meter

5) Initially the **current** through the circuit is **high**. But, as **charge** builds up on the plates, **electrostatic repulsion** makes it **harder** and **harder** for more electrons to be deposited. When the p.d. across the **capacitor** is equal to the p.d. across the **battery**, the **current** falls to **zero**. The capacitor is **fully charged**.

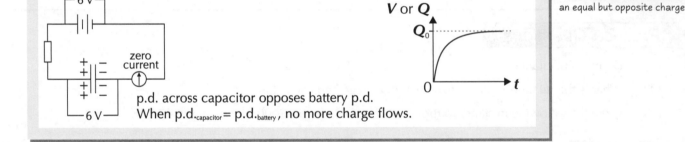

p.d. across capacitor opposes battery p.d.
When p.d.capacitor = p.d.battery, no more charge flows.

an equal but opposite charge

To **Discharge** a Capacitor, **Take Out** the **Battery** and **Reconnect** the **Circuit**

1) When a **charged capacitor** is connected across a **resistor**, the p.d. drives a **current** through the circuit.

2) This current flows in the **opposite direction** from the **charging current**.

3) The capacitor is **fully discharged** when the **p.d.** across the plates and the **current** in the circuit are both **zero**.

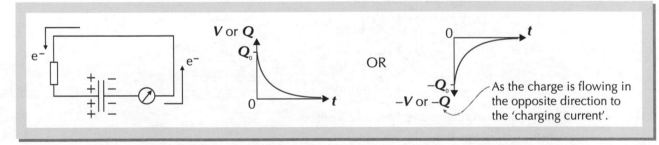

OR

As the charge is flowing in the opposite direction to the 'charging current'.

The **Time Taken** to **Charge** or **Discharge** Depends on **Two Factors**

The **time** it takes to charge up or discharge a capacitor depends on:

1) The **capacitance** of the capacitor (C). This affects the amount of **charge** that can be transferred at a given **voltage**.

2) The **resistance** of the circuit (R). This affects the **current** in the circuit.

Charging and Discharging

The **Charge** on a Capacitor **Decreases Exponentially**

1) When a capacitor is **discharging**, the amount of **charge** left on the plates falls **exponentially with time**.

2) That means it always takes the **same length of time** for the charge to **halve**, no matter **how much charge** you start with — like radioactive decay (see p. 130).

The charge left on the plates of a capacitor discharging from full is given by the equation:

$$Q = Q_0 e^{-\frac{t}{CR}}$$

where Q_0 is the charge of the capacitor when it's fully charged.

The graphs of V against t and I against t for charging and discharging are also exponential.

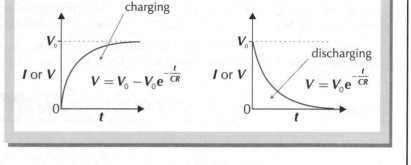

charging

I or V $V = V_0 - V_0 e^{-\frac{t}{CR}}$

discharging

I or V $V = V_0 e^{-\frac{t}{CR}}$

Time Constant τ = CR

τ is the Greek letter 'tau'

If $t = \tau = CR$ is put into the equation above, then $Q = Q_0 e^{-1}$. So when $t = \tau$: $\dfrac{Q}{Q_0} = \dfrac{1}{e}$, where $\dfrac{1}{e} \approx \dfrac{1}{2.718} \approx 0.37$.

1) So τ, the **time constant**, is the time taken for the charge on a discharging capacitor (Q) to **fall** to **37%** of Q_0.

2) It's also the time taken for the charge of a charging capacitor to **rise** to **63%** of Q_0.

3) The **larger** the **resistor** in series with the capacitor, the **longer it takes** to charge or discharge.

4) In practice, the time taken for a capacitor to charge or discharge **fully** is taken to be about 5**CR**.

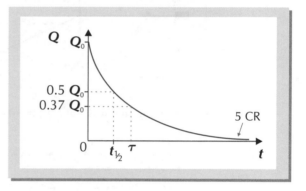

Practice Questions

Q1 Sketch graphs to show the variation of p.d. across the plates of a capacitor with time for:
a) charging a capacitor, b) discharging a capacitor.

Q2 What two factors affect the rate of charge of a capacitor?

Q3 What is meant by the term 'time constant' in relation to capacitors?

Exam Question

Q1 A 250 µF capacitor is fully charged from a 6 V battery and then discharged through a 1 kΩ resistor.

(a) Calculate the time taken for the charge on the capacitor to fall to 37% of its original value. [2 marks]

(b) Calculate the percentage of the total charge remaining on the capacitor after 0.7s. [3 marks]

(c) If the charging voltage is increased to 12 V, what effect will this have on:

  i) the total charge stored

  ii) the capacitance of the capacitor

  iii) the time taken to fully charge [3 marks]

An analogy — consider the lowly bike pump...

A good way to think of the charging process is like pumping air into a bike tyre. To start with, the air goes in easily, but as the pressure in the tyre increases, it gets harder and harder to squeeze any more air in. The tyre's 'full' when the pressure of the air in the tyre equals the pressure of the pump. The analogy works just as well for discharging...

Scattering to Determine Structure

This page covers the history of the atom during the last century — from Thomson's plum pudding to Rutherford's nuclear model and protons and neutrons.

The **Thomson Model** was Popular in the **19th Century**

Until the early 20th century, physicists believed that the atom was a **positively charged globule** with **negatively charged electrons sprinkled** in it. This "**plum pudding**" model of the atom was known as the **Thomson Model**. This all changed in 1909 when the **Rutherford scattering experiment** was done.

In Rutherford's laboratory, **Hans Geiger** and **Ernest Marsden** studied the scattering of **alpha particles** by **thin metal foils**.

Rutherford's Experiment Disproved the Thomson Model

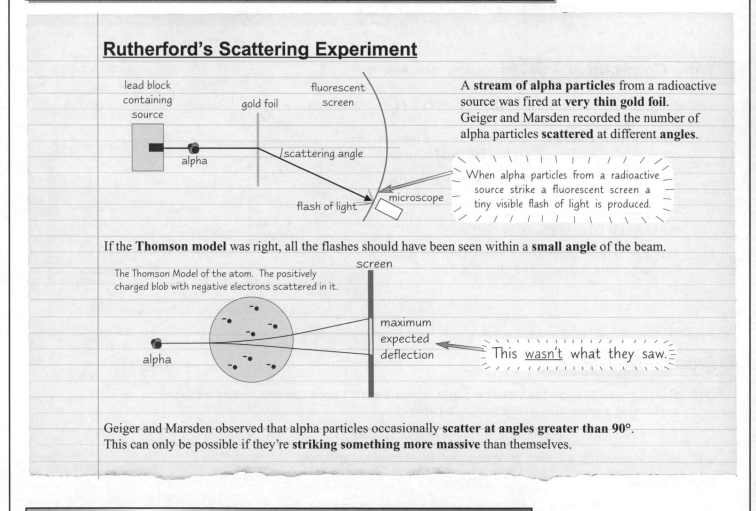

Rutherford's Scattering Experiment

lead block containing source

gold foil

fluorescent screen

alpha

scattering angle

flash of light

microscope

A **stream of alpha particles** from a radioactive source was fired at **very thin gold foil**. Geiger and Marsden recorded the number of alpha particles **scattered** at different **angles**.

When alpha particles from a radioactive source strike a fluorescent screen a tiny visible flash of light is produced.

If the **Thomson model** was right, all the flashes should have been seen within a **small angle** of the beam.

The Thomson Model of the atom. The positively charged blob with negative electrons scattered in it.

alpha

screen

maximum expected deflection

This *wasn't* what they saw.

Geiger and Marsden observed that alpha particles occasionally **scatter at angles greater than 90°**. This can only be possible if they're **striking something more massive** than themselves.

Rutherford's **Model** of the **Atom** — **The Nuclear Model**

This experiment led Rutherford to some **important conclusions**:

1) Most of the **fast, charged alpha particles** went **straight through** the gold foil. Therefore the atom is **mostly empty space**.

2) **Some** of the alpha particles are **deflected back** through **significant angles**, so the **centre** of the atom must be **tiny** but contain **a lot of mass**. Rutherford named this the **nucleus**.

3) The **alpha particles** were **repelled**, so the **nucleus** must have **positive charge**.

4) **Atoms** are **neutral overall** so the **electrons** must be on the outside of the atom — separating one atom from the next.

Scattering to Determine Structure

Atoms are made up of Protons, Neutrons and Electrons

Inside **every atom**, there's a **nucleus** containing **protons** and **neutrons**. **Protons** and **neutrons** are both known as **nucleons**. **Orbiting** this core are the **electrons**. This is called the **nuclear model** of the atom.

Although the **proton** and **neutron** are **2000 times** more **massive** than the **electron**, the nucleus only takes up a **tiny proportion** of the atom. The electrons orbit at relatively **vast distances**.

The nucleus is only one **10 000th the size** of the whole atom — most of the atom is **empty space**.

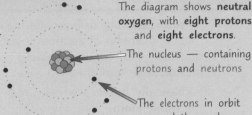

The diagram shows **neutral oxygen**, with **eight protons** and **eight electrons**.

The nucleus — containing protons and neutrons

The electrons in orbit around the nucleus

The Proton Number is the Number of Protons in the Nucleus

No... really.

The **proton number** is sometimes called the **atomic number**, and has the **symbol Z** (don't ask me why). Z is just the **number of protons** in the nucleus.

It's the **proton number** that **defines** the **element** — **no two elements** will have the **same** number of protons.

In a **neutral atom**, the number of **electrons equals** the number of **protons**. The element's **reactions** and **chemical behaviour** depend on the number of **electrons**. So the **proton number** tells you a lot about its **chemical properties**.

The Nucleon Number is the Total Number of Protons and Neutrons

The **nucleon number** is also called the **mass number**, and has the **symbol A** (*shrug*). It tells you how many **protons** and **neutrons** are in the nucleus.

Each **proton or neutron** has a **mass** of (approximately) **1 atomic mass unit** (see p.140). The mass of an electron compared with a nucleon is virtually nothing, so the **number** of **nucleons** is about the same as the **atom's mass** (in atomic mass units).

Atomic Structure can be Represented Using Standard Notation

STANDARD NOTATION:

The **proton number** or **atomic number** (Z) — there are six protons in a carbon atom.

The **nucleon number** or **mass number** (A) — there are a total of 12 protons and neutrons in a carbon-12 atom.

$$^{12}_{6}\text{C}$$

The symbol for the element carbon.

Practice Questions

Q1 List the particles that make up the atom and give their charges and relative masses.

Q2 Define the proton number and nucleon number.

Exam Questions

Q1 In 1911, Ernest Rutherford proposed the nuclear model of the atom after experiments using alpha-particle scattering.

(a) Describe the nuclear model of the atom. [3 marks]

(b) Explain how the alpha-particle scattering experiment provided evidence for Rutherford's model. [3 marks]

Q2 How many protons, neutrons and electrons are there in a (neutral) $^{16}_{8}\text{O}$ atom? [2 marks]

Alpha scattering — it's positively repulsive...

The important things to learn from these two pages are the nuclear model for the structure of the atom (i.e. a large mass nucleus surrounded by orbiting electrons) and how Geiger and Marsden's alpha-particle scattering experiment gives evidence that supports this model. Once you know that, take a deep breath — it's about to get a little more confusing.

Nuclear Radius and Density

The tiny nucleus — such a weird place, but one that you need to become ultra familiar with. Lucky you...

The Nucleus is a Very Small Part of a Whole Atom

1) By **probing atoms** using scattering and diffraction methods, we know that the **diameter of an atom** is about 0.1 nm (1×10^{-10} m) and the diameter of the smallest **nucleus** is about 2 fm (2×10^{-15} m — pronounced "femtometres").

2) So basically, **nuclei** are really, really **tiny** compared with the size of the **whole atom**.

3) To make this **easier to visualise**, try imagining a **large Ferris wheel** (which is pretty darn big) as the size of **an atom**. If you then put a **grain of rice** (which is rather small) in the centre, this would be the size of the atom's **nucleus**.

4) **Molecules** are just a number of **atoms joined together**. As a rough guide, the size of a molecule equals the number of atoms in it multiplied by the size of one atom.

The Nucleus is Made Up of Nucleons

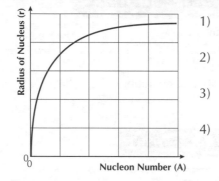

1) The **particles** that make up the nucleus (i.e. **protons** and **neutrons**) are called **nucleons**.

2) The **number of nucleons** in an atom is called the **mass** (or **nucleon**) **number, A**.

3) As **more nucleons** are added to the nucleus, it gets **bigger**.

4) And as we all know by now, you can measure the size of a nucleus by firing particles at it (see p. 122).

See p. 123 for more on the mass number and how this is used to represent atomic structure in standard notation.

Isotopes have the Same Proton Number, but Different Nucleon Numbers

Atoms with the **same number of protons** but **different numbers of neutrons** are called **isotopes**.

Example: Hydrogen has three isotopes — hydrogen, deuterium and tritium
Hydrogen has 1 proton and 0 neutrons.
Deuterium has 1 proton and 1 neutron.
Tritium has 1 proton and 2 neutrons.

Changing the number of **neutrons doesn't affect** the atom's **chemical** properties.

The **number of neutrons** affects the **stability** of the nucleus though (see p. 132).

Unstable nuclei may be **radioactive** (see p. 128).

The Density of Nuclear Matter is Enormous

1) The **volume** that each nucleon (i.e. a **proton** or a **neutron**) takes up in a nucleus is about the **same**.

2) Because protons and neutrons have nearly the **same mass**, it means that all nuclei have a **similar density** (ρ).

3) But nuclear matter is **no ordinary** stuff. Its density is **enormous**. A **teaspoon** of pure nuclear matter would have a mass of about **five hundred million tonnes**. (Just to make you gasp in awe and wonder, out in space nuclear matter makes up neutron stars, which are several kilometres in diameter.)

Nuclear Radius and Density

All Elements Have a Similar Nuclear Density

To calculate **nuclear density** you need to know the **mass** and **volume** of the nucleus — and the equation for density:

$$\rho = \frac{m}{V}$$

The symbol for density, ρ, is the Greek letter 'rho'. Its unit is kg m⁻³.

If you're asked to **estimate nuclear density**, you might have to **work out** the **volume** of the nucleus from its **radius**. So, just assume the nucleus is a sphere and bung the value into the equation:

$$V = \frac{4}{3}\pi r^3$$

Make sure the radius is in m so your volume is in m³.

The following examples show how **nuclear density** is pretty much the **same, regardless of the element**.

Example 1

Work out the density of a carbon nucleus given that has a radius of 3.2×10^{-15} m and a mass of 2.00×10^{-26} kg.

1) **Volume (V) of the nucleus** $= \frac{4}{3}\pi r^3$

$$= 1.37 \times 10^{-43} \text{ m}^3$$

2) This gives the **density (ρ)** of a carbon nucleus as:

$$\rho = \frac{m}{V} = \frac{2.00 \times 10^{-26}}{1.37 \times 10^{-43}} = 1.46 \times 10^{17} \text{ kg m}^{-3}$$

Example 2

Work out the density of a gold nucleus given that it has a radius of 8.1×10^{-15} m and a mass of 3.27×10^{-25} kg.

1) **Volume (V) of the nucleus** $= \frac{4}{3}\pi r^3$

$$= 2.23 \times 10^{-42} \text{ m}^3$$

2) This gives the **density (ρ)** of a gold nucleus as:

$$\rho = \frac{m}{V} = \frac{3.27 \times 10^{-25}}{2.23 \times 10^{-42}} = 1.47 \times 10^{17} \text{ kg m}^{-3}$$

Nuclear density is significantly larger than atomic density — this suggests three important facts about the structure of an atom:
   a) Most of an atom's mass is in its nucleus.
   b) The nucleus is small compared to the atom.
   c) An atom must contain a lot of empty space.

You should expect your estimates of nuclear density to come out around 10¹⁷ kg m⁻³ — if not, there's probably an error in your working.

Practice Questions

Q1   What is the approximate size of an atom?

Q2   What are nucleons?

Q3   What is the term for two atoms with the same number of protons, but different numbers of neutrons?

Q4   Explain why the density of ordinary matter is much less than that of nuclear matter.

Exam Questions

Q1   Define the term 'isotope'. Describe the similarities and differences between the properties of two isotopes of the same element.   [3 marks]

Q2   A radium nucleus has a radius of 8.53×10^{-15} m and a mass of 3.75×10^{-25} kg. Estimate the density of the radium nucleus.   [2 marks]

Q3   A sample of pure gold has a density of 19 300 kg m⁻³. If the density of a gold nucleus is 1.47×10^{17} kg m⁻³, discuss what this implies about the structure of a gold atom.   [4 marks]

Nuclear and particle physics — heavy stuff...

So the nucleus is a tiny part of the atom that's incredibly dense. The density doesn't change much from element to element, and nuclear stability depends on the number of neutrons. Learn the theory like your own backyard, but don't worry about remembering equations and values — those friendly examiners have popped them in the exam paper for you. How nice...

The Strong Nuclear Force

Keeping the nucleus together requires a lot of effort — a bit like A2 Physics then...

There are **Forces** at Work **Inside** the **Nucleus**...

There are several different **forces** acting on the nucleons in a nucleus. To understand these forces you first need to take a look at the **electrostatic** forces due to the protons' electric charges, and also the **gravitational** forces.

1) The **electrostatic force**

All protons have an equal, **positive electric charge**. So, packed close together inside a nucleus, these protons will **repel** each other. You can work out the size of this **force of repulsion** using **Coulomb's law**. Assuming that two protons are 1×10^{-14} m apart, the force of repulsion F_R will be:

$$F_R = \frac{1}{4\pi\varepsilon_o} \frac{Q_1 Q_2}{r^2} = \frac{1}{4\pi(8.85\times10^{-12})} \frac{(1.6\times10^{-19})(1.6\times10^{-19})}{(1\times10^{-14})^2} = 2.3 \text{ N}$$

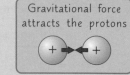

Electrostatic force repels the protons

In this example: Q_1 and Q_2 are the electric charges on two protons, ε_o is the permittivity of free space, and r is the distance separating the two protons.

2) The **gravitational force**

Newton's law of gravitation says that two **massive** objects will **attract** each other. So, for the same two protons, this **attractive force**, F_A, will be:

$$F_A = -G\frac{m_1 m_2}{r^2} = -(6.67\times10^{-11}) \cdot \frac{(1.67\times10^{-27})(1.67\times10^{-27})}{(1\times10^{-14})^2} = -1.86 \times 10^{-36} \text{ N}$$

Gravitational force attracts the protons

In this example: G is the gravitational constant, m_1 and m_2 are the masses of two protons, and r is the distance separating the two protons.

The **electrostatic force** of repulsion is far **bigger** than the **gravitational** attractive force. If these were the only forces acting in the nucleus, the nucleons would **fly apart**. So there must be **another attractive force** that **holds the nucleus together** — called the **strong nuclear force**. (The gravitational force is so small, it's usually ignored.)

The **Strong Nuclear Force** Binds Nucleons Together

Now pay attention please. This bit's rather strange because the **strong nuclear force** is quite **complicated**, but here are the **main points**:

1) To **hold the nucleus together**, the strong nuclear force must be an **attractive force** that is **larger** than the electrostatic force.

2) Experiments have shown that the strong nuclear force between nucleons has a **short range**. It can only hold nucleons together when they are separated by up to **10 fm** — which is the maximum size of a nucleus.

3) The **strength** of the strong nuclear force between nucleons **quickly falls** beyond this distance (see the graph on the next page).

4) Experiments also show that the strong nuclear force **works equally between all nucleons**. This means that the size of the force is the same whether proton-proton, neutron-neutron or proton-neutron.

5) At **very small separations**, the strong nuclear force must be **repulsive** — otherwise there would be nothing to stop it **crushing** the nucleus to a **point**.

lime green, orange and day-glow pink — repulsive at small separations

The Strong Nuclear Force

The **Size** of the Strong Nuclear Force **Varies** with **Nucleon Separation**

The **strong nuclear force** can be plotted on a **graph** to show how it changes with the **distance of separation** between **nucleons**. If the **electrostatic force** is also plotted, you can see the **relationship** between these **two forces**.

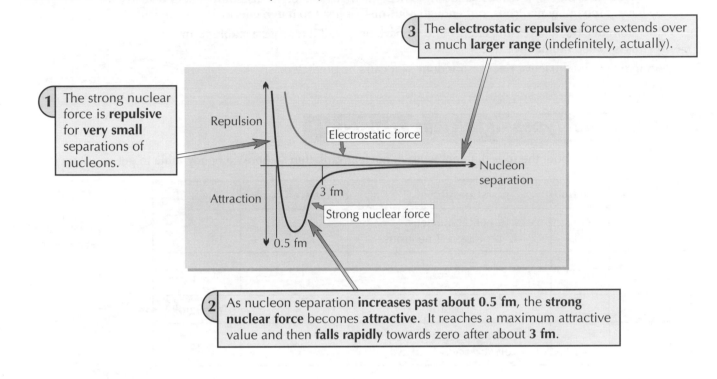

3 The **electrostatic repulsive** force extends over a much **larger range** (indefinitely, actually).

1 The strong nuclear force is **repulsive** for **very small** separations of nucleons.

Repulsion

Electrostatic force

Nucleon separation

Attraction

3 fm

Strong nuclear force

0.5 fm

2 As nucleon separation **increases past about 0.5 fm**, the **strong nuclear force** becomes **attractive**. It reaches a maximum attractive value and then **falls rapidly** towards zero after about **3 fm**.

Practice Questions

Q1 What causes an electrostatic force inside the nucleus?

Q2 Explain why the gravitational forces between nucleons are usually ignored.

Q3 What evidence suggests the existence of a strong nuclear force?

Q4 Is the strong interaction attractive or repulsive at a nucleon separation of 10 fm?

Exam Questions

Q1 Coulomb's law can be used to find the electrostatic force of repulsion between two protons in a nucleus.

$$F = \frac{1}{4\pi\varepsilon_o} \frac{Q_1 Q_2}{r^2}$$

The charge, Q, on a proton is $+1.6 \times 10^{-19}$ C and the permittivity of free space, ε_0, is 8.85×10^{-12} Fm^{-1}.

(a) If two protons are separated by a distance, r, of 9×10^{-15} m, calculate the electrostatic force between them. [2 marks]

(b) If the protons move closer together, what effect will this have on the repulsive force? [1 mark]

(c) What is the electrostatic force between a proton and a neutron? Explain your answer. [2 marks]

Q2 The strong nuclear force binds the nucleus together.

(a) Explain why the force must be repulsive at very short distances. [1 mark]

(b) How does the strong interaction limit the size of a stable nucleus? [2 marks]

The strong interaction's like nuclear glue...

Right then, lots of scary looking stuff on these pages, but DON'T PANIC... the important bits can be condensed into a few points: a) the electrostatic force pushes protons in the nucleus apart, b) the strong interaction pulls all nucleons together, c) there's a point where these forces are balanced — this is the typical nucleon separation in a nucleus. Easy eh?...

Radioactive Emissions

Despite its best intentions, the strong force can't always hold nuclei together — instead you get radioactive emissions.

Unstable Atoms are Radioactive

1) If an atom is **unstable**, it will **break down** to **become** more stable. Its **instability** could be caused by having **too many neutrons**, **not enough neutrons**, or just **too much energy** in the nucleus.

2) The atom **decays** by **releasing energy** and/or **particles**, until it reaches a **stable form** — this is called **radioactive decay**.

3) An individual radioactive decay is **random** — it can't be predicted.

There are Four Types of Nuclear Radiation

You need to know all about the **four** different types of **nuclear radiation** — here's a handy **table** to get you started.

Radiation	Symbol	Constituent	Relative Charge	Mass (u)
Alpha	α	A helium nucleus — 2 protons & 2 neutrons	+2	4
Beta-minus (Beta)	β or β^-	Electron	-1	(negligible)
Beta-plus	β^+	Positron	+1	(negligible)
Gamma	γ	Short-wavelength, high-frequency electromagnetic wave.	0	0

u stands for atomic mass unit — see p. 140.

See p. 136 for more on positrons.

The Different Types of Radiation have Different Penetrations

Alpha, **beta** and **gamma** radiation can be **fired** at a **variety of objects** with **detectors** placed the **other side** to see whether they **penetrate** the object. You can use an **experiment** like this to **identify** a type of radiation.

When a radioactive particle **hits** an **atom** it can **knock off electrons**, creating an **ion** — so, **radioactive emissions** are also known as **ionising radiation**. The **different types** of radiation have **different ionising power** — there's more detail about this on the next page.

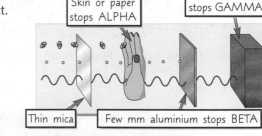

Skin or paper stops ALPHA

Many cm lead stops GAMMA

Thin mica

Few mm aluminium stops BETA

Radiation	Symbol	Ionising	Speed	Penetrating power	Affected by magnetic field
Alpha	α	Strongly	Slow	Absorbed by paper or a few cm of air	Yes
Beta-minus (Beta)	β or β^-	Weakly	Fast	Absorbed by ~3 mm of aluminium	Yes
Beta-plus	β^+	Annihilated by electron — so virtually zero range			
Gamma	γ	Very weakly	Speed of light	Absorbed by many cm of lead, or several m of concrete.	No

Radioactive Emissions

Radioactive Isotopes Have Many Uses

Radioactive substances are extremely useful. You can use them for all sorts — to **date** organic material, diagnose **medical problems**, **sterilise** food, and in **smoke alarms**.

Radiocarbon Dating

The radioactive isotope **carbon-14** is used in **radiocarbon dating**. Living plants take in carbon dioxide from the atmosphere as part of **photosynthesis**, including the **radioactive isotope carbon-14**. When they die, the **activity** of carbon-14 in the plant starts to **fall**, with a **half-life** of around **5730 years**. Archaeological finds made from once living material (like wood) can be tested to find the **current amount** of carbon-14 in them, and date them.

Smoke Detectors

Smoke detectors have a **weak** source of α-**radiation** close to **two electrodes**. The radiation **ionises** the air, and a **current** flows between the electrodes. If there's a fire, **smoke absorbs** the **radiation** — the **current stops** and the **alarm sounds**.

Alpha and Beta Particles have Different Ionising Properties

What a **radioactive source** can be **used** for often depends on its **ionising properties**.

1) **Alpha** particles are **strongly positive** — so they can **easily pull electrons** off atoms, **ionising** them.

2) Ionising an atom **transfers** some of the **energy** from the **alpha particle** to the **atom**. The alpha particle **quickly ionises** many atoms (about 10 000 ionisations per alpha particle) and **loses** all its **energy**. This makes alpha-sources suitable for use in **smoke alarms** because they allow **current** to flow, but won't **travel very far**.

3) The **beta**-minus particle has **lower mass** and **charge** than the alpha particle, but a **higher speed**. This means it can still **knock electrons** off atoms. Each **beta** particle will ionise about 100 atoms, **losing energy** at each interaction.

4) This **lower** number of **interactions** means that beta radiation causes much **less damage** to body tissue than alpha radiation. This means beta radiation can be used in **medicine** to target and damage **cancerous cells** — since it passes through healthy tissue without causing too many problems.

5) Gamma radiation is even more **weakly ionising** than beta radiation, so will do even **less damage** to body tissue. This means it can be used for **diagnostic techniques** in medicine — see pages 150-151 for more info.

Practice Questions

Q1 What makes an atom radioactive?

Q2 What are the four types of nuclear radiation? What does each one consist of?

Q3 Which type of radiation is the most penetrating? Which is the most ionising?

Q4 Suggest two uses of nuclear radiation.

Exam Questions

Q1 Briefly describe an absorption experiment to distinguish between alpha, beta and gamma radiation. You may wish to include a sketch in your answer. [4 marks]

Q2 Radioactive isotopes are found in smoke alarms.

(a) Explain the function of radioactive isotopes in smoke alarms. [3 marks]

(b) What type of radiation is emitted by the isotopes in smoke alarms? Explain why this is suitable. [3 marks]

Radioactive emissions — as easy as α, β, γ...

You need to learn the different types of radiation and their properties. Remember that alpha particles are by far the most ionising and so cause more damage if they get inside your body than the same dose of any other radiation — which is one reason we don't use alpha sources as medical tracers. Learn this all really well, then go and have a brew and a bickie...

Exponential Law of Decay

Oooh look — some maths. Good.

Every Isotope Decays at a Different Rate

1) **Radioactive decay** is completely **random**. You **can't predict which** atom will decay **when**.

2) Although you can't predict the decay of an **individual atom**, if you take a **very large number of atoms**, their **overall behaviour** shows a **pattern**.

3) Any sample of a particular **isotope** (page 124) has the **same rate of decay**, i.e. the same **proportion** of atoms will **decay** in a **given time**.

It could be you.

The Rate of Decay is Measured by the Decay Constant

The **activity** of a sample — the **number** of atoms that **decay each second** — is **proportional** to the **size of the sample**. For a **given isotope**, a sample **twice** as big would give **twice** the **number of decays** per second.

The **decay constant** (λ) measures how **quickly** an isotope will **decay** — the **bigger** the value of λ, the faster the rate of decay. Its unit is s^{-1}.

activity = decay constant × number of atoms

Or in symbols: $A = \lambda N$

Don't get λ confused with wavelength.

Activity is measured in **becquerels** (Bq):   1 Bq = 1 decay per second (s^{-1})

You Need to Learn the Definition of Half-Life

The **half-life** ($T_{1/2}$) of an **isotope** is the **average time** it takes for the **number of undecayed atoms to halve**.

Measuring the **number of undecayed atoms** isn't the easiest job in the world. **In practice**, half-life isn't measured by counting atoms, but by measuring the **time it takes** the **activity** to **halve**.

The **longer** the **half-life** of an isotope, the **longer** it stays **radioactive**.

The Number of Undecayed Particles Decreases Exponentially

You **can't** tell when any **one particle** is going to **decay**, but you can **predict** how many particles are left to decay using a **graph** like the one below. The number of **undecayed** particles of an isotope always **decreases exponentially**.

You can even use the **graph** to find the **half-life**.

How to find the half-life of an isotope

STEP 1: Read off the value of count rate, particles or activity when t = 0.

STEP 2: Go to half the original value.

STEP 3: Draw a horizontal line to the curve, then a vertical line down to the x-axis.

STEP 4: Read off the half-life where the line crosses the x-axis.

STEP 5: Check the units carefully.

STEP 6: It's always a good idea to check your answer. Repeat steps 1-4 for a quarter the original value. Divide your answer by two. That will also give you the half-life. Check that you get the same answer both ways.

The half-life stays the same. It takes the same amount of time for half of the atoms to decay regardless of the number of atoms you start with.

The number of atoms approaches zero.

When you're **measuring** the **activity** and **half-life** of a **source**, you've got to **remember background radiation**.
The **background radiation** needs to be **subtracted** from the **activity readings** to give the **source activity**.

You'd be **more likely** to actually meet a **count rate-time graph** or an **activity-time graph**. They're both **exactly the same shape** as the graph above, but with different **y-axes**.

Exponential Law of Decay

You Need to Know the *Equations* for *Half-Life* and *Decay*...

1) The number of radioactive nuclei decaying per second (**activity**) is proportional to the number of nuclei remaining.

2) The **half-life** can be **calculated** using the equation:
(where ln is the natural log)

$$T_{\frac{1}{2}} = \frac{\ln 2}{\lambda} \simeq \frac{0.693}{\lambda}$$

3) The **number of radioactive atoms** remaining, **N**, depends on the **number originally** present, N_o. The **number remaining** can be calculated using the equation:

$$N = N_0 e^{-\lambda t}$$

Here t = time, measured in seconds.

4) As a sample decays, its activity goes down — there's an equation for that too:

$$A = A_0 e^{-\lambda t}$$

Example:
A sample of the radioactive isotope ^{13}N contains 5×10^6 atoms. The decay constant for this isotope is 1.16×10^{-3} s^{-1}.

a) What is the half-life for this isotope?

$$T_{\frac{1}{2}} = \frac{\ln 2}{1.16 \times 10^{-3}} = 598 \text{ s}$$

b) How many atoms of ^{13}N will remain after 800 seconds?

$$N = N_0 e^{-\lambda t} = 5 \times 10^6 e^{-(1.16 \times 10^{-3})(800)} = 1.98 \times 10^6 \text{ atoms}$$

Capacitors and *Radioactive Isotopes* Have Similar *Decay Equations*

Radioactive isotopes might seem very different from **capacitors** in C-R circuits (see page 121), but their **decay equations** are actually **very similar**. This **table** shows the **similarities** and **differences** between the equations.

	Discharging Capacitors	Radioactive Isotopes
1)	Decay equation is $Q = Q_0 e^{-t/CR}$.	Decay equation is $N = N_0 e^{-\lambda t}$.
2)	The **quantity** that decays is **Q**, the amount of charge left on the plates of the capacitor.	The **quantity** that decays is **N**, the number of unstable nuclei remaining.
3)	**Initially**, the charge on the plates is Q_0.	**Initially**, the number of nuclei is N_0.
4)	It takes **CR** seconds for the amount of charge remaining to fall to **37% of its initial value**.	It takes **1/λ** seconds for the number of nuclei remaining to fall to **37% of the initial value**.
5)	The time taken for the amount of charge left to decay by half (the **half-life**) is $t_{\frac{1}{2}} = \ln 2 \times CR$.	The time taken for the number of nuclei to decay by half (the **half-life**) is $t_{\frac{1}{2}} = \ln 2 / \lambda$.

Practice Questions

Q1 Define radioactive activity. What units is it measured in?

Q2 Sketch a general radioactive decay graph showing the number of particles against time.

Q3 What is meant by the term 'half-life'?

Exam Questions

Q1 Explain what is meant by the random nature of radioactive decay. [1 mark]

Q2 You take a reading of 750 Bq from a pure radioactive source. The radioactive source initially contains 50 000 atoms, and background activity in your lab is measured as 50 Bq.
  (a) Calculate the decay constant for your sample. [3 marks]
  (b) What is the half-life of this sample? [2 marks]
  (c) Approximately how many atoms of the radioactive source will there be after 300 seconds? [2 marks]

Radioactivity is a random process — just like revision shouldn't be...

Remember the shape of that graph — whether it's count rate, activity or number of atoms plotted against time, the shape's always the same. This is all pretty straightforward mathsy-type stuff: plugging values in equations, reading off graphs, etc. Not very interesting, though. Ah well, give it a few pages and you might just be longing for a bit of boredom.

Nuclear Decay

The stuff on these pages covers the most important facts about nuclear decay that you're just going to have to make sure you know inside out. I'd be very surprised if you didn't get a question about it in your exam...

Some Nuclei are **More Stable** than Others

The nucleus is under the **influence** of the **strong nuclear force holding** it **together** and the **electromagnetic force pushing** the **protons apart**. It's a very **delicate balance**, and it's easy for a nucleus to become **unstable**. You can get a stability graph by plotting **Z** (atomic number) against **N** (number of neutrons).

> A nucleus will be **unstable** if it has:
> 1) **too many neutrons**
> 2) **too few neutrons**
> 3) **too many nucleons** altogether, i.e. it's **too heavy**
> 4) **too much energy**

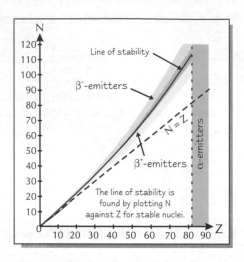

You Can **Represent Nuclear Decay** Using **Equations**

Don't panic — 'nuclear decay equations' sounds a lot more complicated than it really is. You just write down what you **start** off with on the **left**, stick a big **arrow** in the middle, then put the bits you **end** up with on the **right** — easy.

You usually write the particles in **standard notation** (see page 123) so you can see exactly what happens to the **protons** and **neutrons**.

For example, the decay of americium-241 to neptunium-237 looks like this:

$$^{241}_{95}\text{Am} \longrightarrow {}^{237}_{93}\text{Np} + {}^{4}_{2}\alpha$$

α **Emission** Happens in **Heavy Nuclei**

When an alpha particle is **emitted**:

nucleon number decreases by 4

The **proton number decreases** by **two**, and the **nucleon number decreases** by **four**.

$$^{238}_{92}\text{U} \longrightarrow {}^{234}_{90}\text{Th} + {}^{4}_{2}\alpha + {}^{0}_{0}\nu_e$$

proton number decreases by 2

1) **Alpha emission** only happens in **very heavy** atoms (with more than 82 protons), like **uranium** and **radium**.

2) The **nuclei** of these atoms are **too massive** to be stable.

3) An alpha particle and a **neutrino** (ν_e) are emitted (see p. 135).

β^- **Emission** Happens in **Neutron Rich** Nuclei

1) **Beta-minus** (usually just called beta) decay is the emission of an **electron** from the **nucleus** along with an **antineutrino** (see p. 136).

2) Beta decay happens in isotopes that are **"neutron rich"** (i.e. have many more **neutrons** than **protons** in their nucleus).

3) When a nucleus ejects a beta particle, one of the **neutrons** in the nucleus is **changed** into a **proton**.

> In **beta-plus emission**, a **proton** gets **changed** into a **neutron**. The **proton number decreases** by **one**, and the **nucleon number stays the same**.

When a beta-minus particle is **emitted**:

nucleon number stays the same

The **proton number increases** by **one**, and the **nucleon number stays the same**.

$$^{188}_{75}\text{Re} \longrightarrow {}^{188}_{76}\text{Os} + {}^{0}_{-1}\beta + {}^{0}_{0}\bar{\nu}_e$$

proton number increases by 1

Nuclear Decay

γ Radiation is Emitted from Nuclei with Too Much Energy

Gamma rays can be emitted from a nucleus with **excess energy** — it's **excited**. This energy is **lost** by emitting a **gamma ray**. This often happens after an **alpha** or **beta** decay.

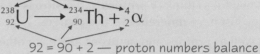

The artificial isotope technetium-99ᵐ is formed in an excited state. It is used as a tracer in medical imaging (see p 156).

> During **gamma emission**, there is **no change** to the nuclear **constituents** — the nucleus just **loses excess energy**.

There are Conservation Rules in Nuclear Reactions

In every nuclear reaction **energy**, **momentum**, **proton number / charge** and **nucleon number** must be conserved.

For example, if a **nucleus** has **92 protons** before it decays, then the **decay products** must have **92 protons** between them — it's that simple.

$$238 = 234 + 4 \text{ — nucleon numbers balance}$$
$$^{238}_{92}U \longrightarrow \,^{234}_{90}Th + \,^{4}_{2}\alpha$$
$$92 = 90 + 2 \text{ — proton numbers balance}$$

Mass is Not Conserved

1) The **mass** of the **alpha particle** is less than the **individual masses** of **two protons** and **two neutrons**. The difference is called the **mass defect** (see page 140).

2) Mass **doesn't** have to be **conserved** because of **Einstein's equation**: $E = mc^2$

3) This says that **mass and energy** are **equivalent**. The **energy released** when the nucleons **bonded together** accounts for the missing mass — so the **energy released** is the same as the **mass defect × c^2**.

Cecil was proud of his ability to conserve mass.

Practice Questions

Q1 What makes a nucleus unstable? Describe the changes that happen in the nucleus during alpha, beta and gamma decay.

Q2 Explain the circumstances in which gamma radiation may be emitted.

Q3 Define the mass defect.

Exam Questions

Q1 (a) Radium-226 undergoes alpha decay to radon. Complete the balanced nuclear equation for this reaction.

$$^{226}_{88}Ra \rightarrow \quad Rn +$$

[3 marks]

(b) Potassium-40 ($Z = 19$, $A = 40$) undergoes beta decay to calcium. Write a balanced nuclear equation for this reaction.

[3 marks]

Q2 Calculate the energy released during the formation of an alpha particle, given that the total mass of two protons and two neutrons is 6.695×10^{-27} kg, the mass of an alpha particle is 6.645×10^{-27} kg and the speed of light, c, is 3.00×10^8 ms⁻¹.

[3 marks]

Nuclear decay — it can be enough to make you unstable...

$E = mc^2$ is an important equation that says mass and energy are equivalent. Remember it well, 'cos you're going to come across it a lot in questions about mass defect and the energy released in nuclear reactions over the next few pages...

Classification of Particles

There are loads of different types of particle apart from the ones you get in normal matter (protons, neutrons, etc.).
They only appear in cosmic rays and in particle accelerators, and they often decay very quickly, so they're difficult to get
a handle on. Nonetheless, you need to learn about a load of them and their properties.

Don't expect to really understand this (I don't) — you only need to learn it. Stick with it — you'll get there.

Hadrons are Particles that Feel the Strong Interaction (e.g. Protons and Neutrons)

1) The **nucleus** of an atom is made up from **protons** and **neutrons** held together by the **strong nuclear force** (sometimes called the strong interaction) — see p. 126.

2) **Not all particles** can **feel** the **strong interaction** — the ones that **can** are called **hadrons**.

3) Hadrons aren't **fundamental** particles. They're made up of **smaller particles** called **quarks** (see page 138).

4) There are **two** types of **hadron** — **baryons** and **mesons**.

Protons and Neutrons are Baryons

1) It's helpful to think of **protons** and **neutrons** as **two versions** of the **same particle** — the **nucleon**. They just have **different electric charges**.

2) As well as **protons** and **neutrons**, there are **other baryons** that you don't get in normal matter — like **sigmas** (Σ) — they're **short-lived** and you **don't** need to **know about them** for A2 (woohoo!).

The Proton is the Only Stable Baryon

All baryons except protons decay to a **proton**. Most physicists think that protons don't **decay**.

Some theories predict that protons should decay with a very long half-life of about 10^{32} years — but there's no experimental evidence for it at the moment.

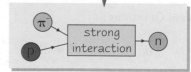

Baryon and Meson felt the strong interaction.

The Number of Baryons in a reaction is called the Baryon Number

Baryon number is the number of baryons. (A bit like **nucleon number** but including unusual baryons like Σ too.) The **proton** and the **neutron** each have a baryon number $B = +1$.

The **total baryon number** in **any** particle reaction **never changes**.

The Mesons You Need to Know About are Pions and Kaons

1) **All mesons** are **unstable** and have **baryon number** $B = 0$ (because they're not baryons).

2) **Pions** (π-mesons) are the **lightest mesons**. You get **three versions** with different **electric charges** — π^+, π^0 and π^-. Pions were **discovered** in **cosmic rays**. You get **loads** of them in **high energy particle collisions** like those studied at the **CERN** particle accelerator.

3) **Kaons** (**K**-mesons) are **heavier** and more **unstable** than **pions**. You get different ones like K^+, K^- and K^0.

4) Mesons **interact** with **baryons** via the **strong interaction**.

Pion interactions swap p's with n's and n's with p's, but leave the overall baryon number unchanged.

π^- → strong interaction → n
p →

Summary of Hadron Properties

DON'T PANIC if you don't understand

all this yet. For now, just **learn** these properties. You'll need to work through to the end of page 139 to see how it **all fits in**.

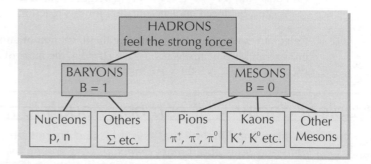

Classification of Particles

Leptons Don't feel the Strong Interaction (e.g. Electrons and Neutrinos)

1) **Leptons** are **fundamental particles** and they **don't** feel the **strong interaction**.
   The only way they can **interact** with other particles is via the **weak interaction**
   and gravity (and the electromagnetic force as well if they're charged).

2) **Electrons** (e^-) are **stable** and very **familiar** but — you guessed it — there are also
   two more leptons called the **muon** (μ^-) and the **tau** (τ^-) that are just like **heavy electrons**.

3) **Muons** and **taus** are **unstable**, and **decay** eventually into **ordinary electrons**.

4) The **electron**, **muon** and **tau** each come with their **own neutrino**: ν_e, ν_μ and ν_τ.

   ν is the Greek letter "nu".

5) **Neutrinos** have **zero** or **almost zero mass** and **zero electric charge** — so they don't do much.
   Neutrinos only take part in **weak interactions** (see p. 139). In fact, a neutrino can **pass right
   through the Earth** without **anything** happening to it.

You Have to Count the Three Types of Lepton Separately

Each lepton is given a **lepton number** of **+1**, but the **electron, muon** and **tau** types of lepton have to be **counted separately**.

You get **three different** lepton numbers: L_e, L_μ and L_τ.

Like the baryon number, the lepton number is just the number of leptons.

Name	Symbol	Charge	L_e	L_μ	L_τ
electron	e^-	-1	$+1$	0	0
electron neutrino	ν_e	0	$+1$	0	0
muon	μ^-	-1	0	$+1$	0
muon neutrino	ν_μ	0	0	$+1$	0
tau	τ^-	-1	0	0	$+1$
tau neutrino	ν_τ	0	0	0	$+1$

Neutrons Decay into Protons

The **neutron** is an **unstable particle** that **decays** into a **proton**. (But it's much more stable when it's part of a nucleus.) It's really just an **example** of β^- decay, which is caused by the **weak interaction**.

$$n \rightarrow p + e^- + \bar{\nu}_e$$

Free neutrons (i.e. ones not held in a nucleus) have a half-life of about 15 minutes.

The antineutrino has $L_e = -1$ so the total lepton number is zero. Antineutrino? Yes, well I haven't mentioned antiparticles yet. Just wait for the next page ...

Practice Questions

Q1 List the differences between a hadron and a lepton.

Q2 Which is the only stable baryon?

Q3 A particle collision at CERN produces 2 protons, 3 pions and 1 neutron.
   What is the total baryon number of these particles?

Q4 Which two particles have lepton number $L_\tau = +1$?

Exam Questions

Q1 List all the decay products of the neutron. Explain why this decay cannot be due to the strong interaction. [3 marks]

Q2 Initially the muon was incorrectly identified as a meson. Explain why the muon is not a meson. [3 marks]

Go back to the top of page 134 — do not pass GO, do not collect £200...

Do it. Go back and read it again. I promise — read these pages about 3 or 4 times and you'll start to see a pattern. There are hadrons that feel the force, leptons that don't. Hadrons are either baryons or mesons, and they're all weird except for those well-known baryons: protons and neutrons. There are loads of leptons, including good old electrons.

Antiparticles

More stuff that seems to laugh in the face of common sense — but actually, antiparticles help to explain a lot in particle physics... (Oh, and if you haven't read pages 134 and 135 yet then go back and read them now — no excuses, off you go...)

Antiparticles were Predicted Before they were Discovered

When **Paul Dirac** wrote down an equation obeyed by **electrons**, he found a kind of **mirror image** solution.

1) It predicted the existence of a particle like the **electron** but with **opposite electric charge** — the **positron**.

2) The **positron** turned up later in a cosmic ray experiment. Positrons are **antileptons** so $L_e = -1$ for them. They have **identical mass** to electrons but they carry a **positive** charge.

Nice one Paul. Now we've got twice as many things to worry about.

Every Particle has an Antiparticle

Each particle type has a **corresponding antiparticle** with the **same mass** but with **opposite charge**. For instance, an **antiproton** is a **negatively charged** particle with the same mass as the **proton**.

Even the shadowy **neutrino** has an antiparticle version called the **antineutrino** — it doesn't do much either.

Particle	Symbol	Charge	B	L_e	Antiparticle	Symbol	Charge	B	L_e
proton	p	+1	+1	0	antiproton	$\bar{p}$	–1	–1	0
neutron	n	0	+1	0	antineutron	$\bar{n}$	0	–1	0
electron	e	–1	0	+1	positron	e^+	+1	0	–1
electron neutrino	ν_e	0	0	+1	electron antineutrino	$\bar{\nu}_e$	0	0	–1

You can Create Matter and Antimatter from Energy

You've probably heard about the **equivalence** of energy and mass. It all comes out of Einstein's special theory of relativity. **Energy** can turn into **mass** and **mass** can turn into **energy** if you know how — all you need is one fantastic and rather famous formula. ⟹

$$E = mc^2$$

It's a good thing this doesn't randomly happen all the time or else you could end up with cute bunny rabbits popping up and exploding unexpectedly all over the place. Oh, the horror...

As you've probably guessed, there's a bit **more to it** than that:

When **energy** is converted into **mass** you have to make **equal amounts** of **matter** and **antimatter**.

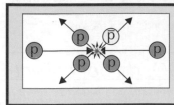

Fire **two protons** at each other at high speed and you'll end up with a lot of **energy** at the point of impact. This energy can form **more particles**.

If an extra **proton** is created, there has to be an **antiproton** made to go with it. It's called **pair production**.

Antiparticles

Each **Particle-Antiparticle Pair** is Produced from a **Single Photon**

Pair production only happens if **one gamma ray photon** has enough energy to produce that much mass. It also tends to happen near a **nucleus**, which helps conserve momentum.

You usually get **electron-positron** pairs produced (rather than any other pair) — because they have a relatively **low mass**.

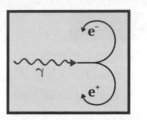

The particle tracks are curved because there's usually a magnetic field present in particle physics experiments. They curve in opposite directions because of the opposite charges on the electron and positron.

The **Opposite** of **Pair Production** is **Annihilation**

When a **particle** meets its **antiparticle** the result is **annihilation**. All the **mass** of the particle and antiparticle gets converted to **energy**. In ordinary matter antiparticles can only exist for a fraction of a second before this happens, so you won't see many of them.

The electron and positron annihilate and their mass is converted into the energy of a pair of gamma ray photons.

Mesons are Their **Own Antiparticles**

Just before you leave this bit it's worth mentioning that the π^- meson is just the **antiparticle** of the π^+ meson, and the **antiparticle** of a π^0 meson is **itself**. You'll see why on p. 138. So we don't need any more particles here... Phew.

(If you don't know what a meson is, look back at page 134.)

Practice Questions

Q1 Which antiparticle has zero charge and a baryon number of –1?

Q2 Describe the properties of an electron antineutrino.

Q3 What is pair production? What happens when a proton collides with an antiproton?

Exam Questions

Q1 Write down an equation for the reaction between a positron and an electron and give the name for this type of reaction. [2 marks]

Q2 According to Einstein, mass and energy are equivalent. Explain why the mass of a block of iron cannot be converted directly into energy. [2 marks]

Q3 Give a reason why the reaction $p + p \rightarrow p + p + n$ is not possible. [1 mark]

Now stop meson around and do some work...

The idea of every particle having an antiparticle might seem a bit strange, but just make sure you know the main points — a) if energy is converted into a particle, you also get an antiparticle, b) an antiparticle won't last long before it bumps into the right particle and annihilates it with a big ba-da-boom, c) this releases the energy it took to make them to start with...

Quarks

*If you haven't read pages 134 to 137, do it now! For the rest of you — here are the **juicy bits** you've been waiting for. Particle physics makes **a lot more sense** when you look at quarks. More sense than it did before anyway.*

Quarks are Fundamental Particles

Quarks are the **building blocks** for **hadrons** (baryons and mesons). *If that first sentence doesn't make much sense to you, read pages 134-137 — you have been warned... twice.*

1) To make **protons** and **neutrons** you only need two types of quark — the **up** quark (**u**) and the **down** quark (**d**).

2) An extra one called the **strange** quark (**s**) lets you make more particles with a property called **strangeness**.

3) There are another three types of quark called **top**, **bottom** and **charm** (tut... physicists) that were predicted from the symmetry of the quark model. But luckily you don't have to know much about them...

The antiparticles of hadrons are made from **antiquarks**.

Quarks and Antiquarks have Opposite Properties

The **antiquarks** have **opposite properties** to the quarks — as you'd expect.

QUARKS

name	symbol	charge	baryon number	strangeness
up	u	$+\frac{2}{3}$	$+\frac{1}{3}$	0
down	d	$-\frac{1}{3}$	$+\frac{1}{3}$	0
strange	s	$-\frac{1}{3}$	$+\frac{1}{3}$	−1

ANTIQUARKS

name	symbol	charge	baryon number	strangeness
anti-up	$\bar{u}$	$-\frac{2}{3}$	$-\frac{1}{3}$	0
anti-down	$\bar{d}$	$+\frac{1}{3}$	$-\frac{1}{3}$	0
anti-strange	$\bar{s}$	$+\frac{1}{3}$	$-\frac{1}{3}$	+1

Baryons are Made from Three Quarks

Evidence for quarks came from **hitting protons** with **high energy electrons**.
The way the **electrons scattered** showed that there were **three concentrations of charge** (quarks) **inside** the proton.

Proton = **uud**

Total charge
= 2/3 + 2/3 − 1/3 = 1
Baryon number
= 1/3 + 1/3 + 1/3 = 1

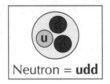

Neutron = **udd**

Total charge
= 2/3 − 1/3 − 1/3 = 0
Baryon number
= 1/3 + 1/3 + 1/3 = 1

Antiprotons are $\bar{u}\bar{u}\bar{d}$ and antineutrons are $\bar{u}\bar{d}\bar{d}$ — so no surprises there then.

Mesons are a Quark and an Antiquark

Pions are just made from **up** and **down** quarks and their **antiquarks**.
Kaons have **strangeness** so you need to put in **s** quarks as well
(remember that the **s** quark has a strangeness of S = −1).

Physicists love patterns. Gaps in patterns like this predicted the existence of particles that were actually found later in experiments. Great stuff.

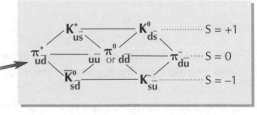

There's no Such Thing as a Free Quark

What if you **blasted** a **proton** with **enough energy** — could you **separate out** the quarks? Nope. The energy just gets changed into more **quarks and antiquarks** — it's **pair production** again and you just make **mesons**. This is called **quark confinement**.

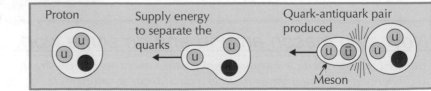

Quarks

The **Weak Interaction** is something that Changes the **Quark Type**

In β⁻ decay a **neutron** is changed into a **proton** — in other words **udd** changes into **uud**.
It means turning a **d** quark into a **u** quark. Only the weak interaction can do this.

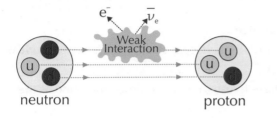

neutron   proton

Some unstable isotopes like **carbon-11** decay by β⁺ emission. In this case a **proton** changes to a **neutron**, so a **u** quark changes to a **d** quark and we get:

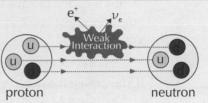

proton   neutron

Four Properties are **Conserved** in **Particle Reactions**

In **any** particle reaction, the **total charge** after the reaction must equal the total charge before the reaction. The same goes for **baryon number**.

The **only** way to change the **type** of quark is with the **weak interaction**, so in strong interactions there has to be the same number of strange quarks at the beginning as at the end.

The reaction $K^- + p \rightarrow n + \pi^0$ is fine for **charge** and **baryon number** but not for **strangeness** — so it won't happen. The negative kaon has an **s** quark in it.

The **three types** of lepton number have to be conserved **separately**.

1) For example, the reaction
$\pi^- \rightarrow \mu^- + \overline{\nu}_\mu$ has $L_\mu = 0$ at the start and $L_\mu = 1 - 1 = 0$ at the end, so it's OK.

2) On the other hand, the reaction $\nu_\mu + \mu^- \rightarrow e^- + \nu_e$ can't happen. At the start $L_\mu = 2$ and $L_e = 0$ but at the end $L_\mu = 0$ and $L_e = 2$.

Sid had been conserving his strangeness for years...

Practice Questions

Q1 What is a quark?

Q2 Which type of particle is made from a quark and an antiquark?

Q3 Describe how a neutron is made up from quarks.

Q4 Explain why quarks are never observed on their own.

Q5 List four quantities that are conserved in particle reactions.

Exam Questions

Q1 State the combination of three quarks that make up a proton. [1 mark]

Q2 Give the quark composition of the π⁻ and explain how the charges of the quarks give rise to its charge. [2 marks]

Q3 Explain how the quark composition is changed in the β⁻ decay of the neutron. [2 marks]

Q4 Give two reasons why the reaction $p + p \rightarrow p + K^+$ does not happen. [2 marks]

A physical property called strangeness — how cool is that...

True, there's a lot of information here, but this page really does tie up a lot of the stuff on the last few pages. Learn as much as you can from this double page, then go back to page 134, and work back through to here. Don't expect to understand it all — but you will definitely find it much easier to learn when you can see how all the bits fit in together.

Binding Energy

Turn off the radio and close the door, 'cos you're going to need to concentrate hard on this stuff about binding energy...

The **Mass Defect** is **Equivalent** to the **Binding Energy**

1) The **mass** of a **nucleus** is **less than** the mass of its **constituent parts** — this is called the **mass defect** (see p. 133).

2) Einstein's equation, $E = mc^2$, says that mass and energy are **equivalent**.

3) So, as nucleons join together, the total mass **decreases** — this 'lost' mass is **converted** into energy and **released**.

4) The amount of **energy released** is **equivalent** to the **mass defect**.

5) If you **pulled** the nucleus completely **apart**, the **energy** you'd have to use to do it would be the **same** as the energy **released** when the nucleus formed.

> The energy needed to **separate** all of the nucleons in a nucleus is called the **binding energy** (measured in **MeV**), and it is **equivalent** to the **mass defect**.

> **Example** Calculate the binding energy of the nucleus of a lithium atom, $_3^6\text{Li}$, given that its mass defect is 0.0343 u.
>
> 1) Convert the mass defect into kg.
>
> $$\text{Mass defect} = 0.0343 \times 1.66 \times 10^{-27} = 5.70 \times 10^{-29} \text{ kg}$$
>
> 2) Use $E = mc^2$ to calculate the binding energy.
>
> $$E = 5.70 \times 10^{-29} \times (3 \times 10^8)^2 = 5.13 \times 10^{-12} \text{ J} = 32 \text{ MeV}$$

Atomic mass is usually given in atomic mass units (u), where $1\text{ u} = 1.66 \times 10^{-27}$ kg.

$1\text{ MeV} = 1.6 \times 10^{-13}$ J

6) The **binding energy per unit of mass defect** can be calculated (using the example above):

$$\frac{\text{binding energy}}{\text{mass defect}} = \frac{32 \text{ MeV}}{0.0343 \text{ u}} \approx 931 \text{ MeVu}^{-1}$$

7) This means that a mass defect of **1 u** is equivalent to about **931 MeV** of binding energy.

Captain Skip didn't believe in ghosts, marmalade and that things could be bound without rope.

The **Binding Energy Per Nucleon** is at a **Maximum** around **N = 50**

A useful way of **comparing** the binding energies of different nuclei is to look at the **binding energy per nucleon**.

> Binding energy per nucleon (in MeV) = $\dfrac{\text{Binding energy (B)}}{\text{Nucleon number (A)}}$

So, the binding energy per nucleon for $_3^6\text{Li}$ (in the example above) is $32 \div 6 = 5.3$ MeV.

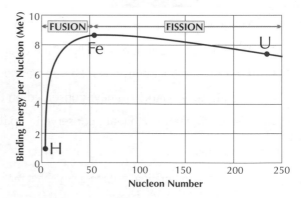

1) A **graph** of **binding energy per nucleon** against **nucleon number**, for all elements, shows a **curve**.

2) **High** binding energy per nucleon means that **more energy** is needed to **remove** nucleons from the nucleus.

3) In other words the **most stable** nuclei occur around the **maximum point** on the graph — which is at **nucleon number 56** (i.e. **iron**, Fe).

4) **Combining small nuclei** is called nuclear **fusion** (see p. 143) — this **increases** the **binding energy per nucleon** dramatically, which means a lot of **energy is released** during nuclear fusion.

5) **Fission** is when **large nuclei** are **split in two** (see p. 142) — the **nucleon numbers** of the two **new nuclei** are **smaller** than the original nucleus, which means there is an **increase** in the binding energy per nucleon. So, energy is also **released** during nuclear fission (but not as much energy per nucleon as in nuclear fusion).

Binding Energy

The **Change** in **Binding Energy** Gives the **Energy Released**...

The **binding energy per nucleon graph** can be used to **estimate** the **energy released** from nuclear reactions.

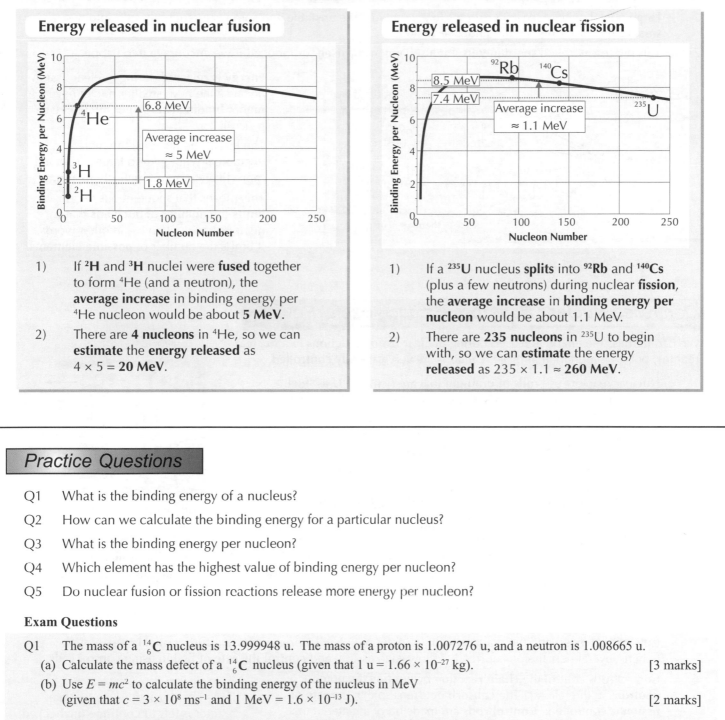

Energy released in nuclear fusion

1) If **²H** and **³H** nuclei were **fused** together to form **⁴He** (and a neutron), the **average increase** in binding energy per ⁴He nucleon would be about **5 MeV**.

2) There are **4 nucleons** in ⁴He, so we can **estimate** the **energy released** as 4 × 5 = **20 MeV**.

Energy released in nuclear fission

1) If a **²³⁵U** nucleus **splits** into **⁹²Rb** and **¹⁴⁰Cs** (plus a few neutrons) during nuclear **fission**, the **average increase** in **binding energy per nucleon** would be about 1.1 MeV.

2) There are **235 nucleons** in ²³⁵U to begin with, so we can **estimate** the energy **released** as 235 × 1.1 ≈ **260 MeV**.

Practice Questions

Q1 What is the binding energy of a nucleus?

Q2 How can we calculate the binding energy for a particular nucleus?

Q3 What is the binding energy per nucleon?

Q4 Which element has the highest value of binding energy per nucleon?

Q5 Do nuclear fusion or fission reactions release more energy per nucleon?

Exam Questions

Q1 The mass of a $^{14}_{6}C$ nucleus is 13.999948 u. The mass of a proton is 1.007276 u, and a neutron is 1.008665 u.
(a) Calculate the mass defect of a $^{14}_{6}C$ nucleus (given that 1 u = 1.66×10^{-27} kg). [3 marks]
(b) Use $E = mc^2$ to calculate the binding energy of the nucleus in MeV (given that $c = 3 \times 10^8$ ms⁻¹ and 1 MeV = 1.6×10^{-13} J). [2 marks]

Q2 The following equation represents a nuclear reaction that takes place in the Sun:
$$^1_1p + ^1_1p \rightarrow ^2_1H + ^0_{+1}\beta + \text{energy released}$$
    where p is a proton and β is a positron (opposite of an electron)
(a) What type of nuclear reaction is this? [1 mark]
(b) Given that the binding energy per nucleon for a proton is 0 MeV and for a ²H nucleus it is approximately 0.86 MeV, estimate the energy released by this reaction. [2 marks]

A mass defect of 1 u is equivalent to a binding energy of 931 MeV...

Remember this useful little fact, and it'll save loads of time in the exam — because you won't have to fiddle around with converting atomic mass from u → kg and binding energy from J → MeV. What more could you possibly want...

Nuclear Fission and Fusion

What did the nuclear scientist have for his tea? Fission chips... hohoho.

Fission *Means* Splitting Up *into* Smaller Parts

1) **Large nuclei**, with at least 83 protons (e.g. uranium), are **unstable** and some can randomly **split** into two **smaller** nuclei — this is called **nuclear fission**.

2) This process is called **spontaneous** if it just happens **by itself**, or **induced** if we **encourage** it to happen.

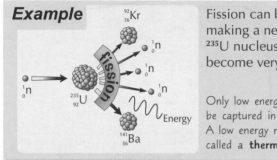

Example

Fission can be induced by making a neutron enter a ^{235}U nucleus, causing it to become very unstable.

Only low energy neutrons can be captured in this way. A low energy neutron is called a **thermal neutron**.

3) **Energy is released** during nuclear fission because the new, smaller nuclei have a **higher binding energy per nucleon** (see p. 140).

4) The **larger** the nucleus, the more **unstable** it will be — so large nuclei are **more likely** to **spontaneously fission**.

5) This means that spontaneous fission **limits** the **number of nucleons** that a nucleus can contain — in other words, it **limits** the number of **possible elements**.

Controlled Nuclear Reactors *Produce Useful* Power

We can **harness** the **energy** released during nuclear **fission reactions** in a **nuclear reactor**, but it's important that these reactions are very **carefully controlled**.

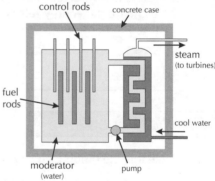

1) Nuclear reactors use **rods of uranium** that are rich in ^{235}U as 'fuel' for fission reactions. (The rods also contain a lot of ^{238}U, but that doesn't undergo fission.)

2) These **fission** reactions produce more **neutrons** which then **induce** other nuclei to fission — this is called a **chain reaction**.

3) The **neutrons** will only cause a chain reaction if they are **slowed down**, which allows them to be **captured** by the uranium nuclei — these slowed down neutrons are called **thermal neutrons**.

4) ^{235}U **fuel rods** need to be placed in a **moderator** (for example, **water**) to **slow down** and/or absorb **neutrons**. You need to choose a moderator that will slow down some neutrons enough so they can cause **further fission**, keeping the reaction going at a steady rate. Choosing a moderator that absorbs **more neutrons the higher the temperature** will **decrease** the chance of **meltdown** if the reactor overheats — as it will naturally **slow down** the reaction.

5) You want the chain reaction to continue on its own at a **steady rate**, where **one** fission follows another. The amount of 'fuel' you need to do this is called the **critical mass** — any less than the critical mass (**sub-critical mass**) and the reaction will just peter out. Nuclear reactors use a **supercritical** mass of fuel (where several new fissions normally follow each fission) and **control the rate of fission** using **control rods**.

6) Control rods control the **chain reaction** by **limiting** the number of **neutrons** in the reactor. They **absorb neutrons** so that the **rate of fission** is controlled. **Control rods** are made up of a material that **absorbs neutrons** (e.g. boron), and they can be inserted by varying amounts to control the reaction rate.
In an **emergency**, the reactor will be **shut down** automatically by the **release of the control rods** into the reactor, which will stop the reaction as quickly as possible.

7) **Coolant** is sent around the reactor to **remove heat** produced in the fission — often the coolant is the **same water** that is being used in the reactor as a **moderator**. The **heat** from the reactor can then be used to make **steam** for powering **electricity-generating turbines**.

If the chain reaction in a nuclear reactor is **left to continue unchecked**, large amounts of **energy** are **released** in a very **short time**.

Many new fissions will follow each fission, causing a **runaway reaction** which could lead to an **explosion**. This is what happens in a **fission (atomic) bomb**.

Nuclear Fission and Fusion

Waste Products of Fission Must be Disposed of Carefully

1) The **waste products** of **nuclear fission** usually have a **larger proportion of neutrons** than stable nuclei of a similar atomic number — this makes them **unstable** and **radioactive**.

2) The products can be used for **practical applications** such as **tracers** in medical diagnosis (see p. 150).

3) However, they may be **highly radioactive** and so their **handling** and **disposal** needs **great care**.

4) When material is removed from the reactor, it is initially **very hot**, so is placed in **cooling ponds** until the **temperature falls** to a safe level.

5) The radioactive waste is then **stored** underground in **sealed containers** until its **activity has fallen** sufficiently.

Fusion Means Joining Nuclei Together

1) **Two light nuclei** can **combine** to create a larger nucleus — this is called **nuclear fusion**.

2) Nuclei can **only fuse** if they have enough energy to overcome the **electrostatic repulsive** force between them, and get close enough for the **strong interaction** to bind them.

3) Typically they need about **1 MeV** of kinetic energy — and that's **a lot of energy**.

Example

In the Sun, **hydrogen nuclei** fuse in a series of reactions to form **helium**.

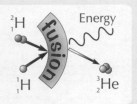

$$^2_1H + {}^1_1H \rightarrow {}^3_2He + \text{energy}$$

Fusion Happens in the Core of Stars

1) The **energy** emitted by the **Sun** and other stars comes from nuclear **fusion** reactions.

2) Fusion can happen because the **temperature** in the **core of stars** is so **high** — the core of the Sun is about 10^7 K.

3) At these temperatures, **atoms don't exist** — the negatively charged electrons are **stripped away**, leaving **positively charged nuclei** and **free electrons**. The resulting mixture is called a **plasma**.

4) A lot of **energy** is released during nuclear fusion because the new, heavier nuclei have a **much higher binding energy per nucleon** (see p. 140). This helps to **maintain the temperature** for further fusion reactions to happen.

5) Experimental **fusion reactors** like JET (the Joint European Torus) are trying to **recreate** these conditions to generate **electricity** (without all the nasty waste you get from fission reactors). Unfortunately, the electricity generated at the moment is **less** than the amount needed to get the reactor up to temperature. But watch this space...

Practice Questions

Q1 What is spontaneous fission?

Q2 How can fission be induced in ^{235}U?

Q3 Why must the waste products of nuclear fission be disposed of very carefully?

Q4 Describe the conditions in the core of a star.

Exam Questions

Q1 Nuclear reactors use carefully controlled chain reactions to produce energy.
   (a) Explain what is meant by the expression 'chain reaction' in terms of nuclear fission.   [2 marks]
   (b) Describe and explain one feature of a nuclear reactor whose role is to control the rate of fission.
      Include an example of a suitable material for the feature you have chosen.   [3 marks]
   (c) Explain what happens in a nuclear reactor during an emergency shut-down.   [2 marks]

Q2 Discuss two advantages and two disadvantages of using nuclear fission to produce electricity.   [4 marks]

If anyone asks, I've gone fission... that joke never gets old...

So, controlled nuclear fission reactions can provide a shedload of energy to generate electricity. There are pros and cons to using fission reactors. They produce huge amounts of energy without so much greenhouse gas, but they leave behind some very nasty radioactive waste... But then, you already knew that — now you need to learn all the grisly details.

X-Ray Imaging

X-ray imaging is one kind of non-invasive diagnostic technique — these techniques let doctors see what's going on (or going wrong) inside your body, without having to open you up and have a look.

X-rays are Produced by Bombarding Tungsten with High Energy Electrons

1) In an X-ray tube, **electrons** are emitted from a **heated filament** and **accelerated** through a high **potential difference** (the **tube voltage**) towards a **tungsten anode**.

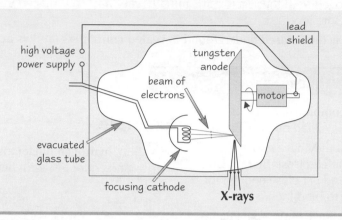

2) When the **electrons** smash into the **tungsten anode**, they **decelerate** and some of their **kinetic energy** is converted into **electromagnetic energy**, as **X-ray photons**. The tungsten anode emits a **continuous spectrum** of **X-ray radiation**.

3)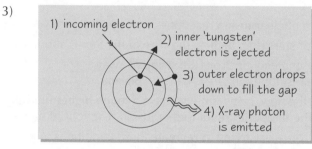

X-rays are also produced when beam electrons **knock out** other electrons from the **inner shells** of the **tungsten atoms**.

Electrons in the atoms' **outer shells** move into the **vacancies** in the **lower energy levels**, and **release energy** in the form of **X-ray photons**.

4) Only about **1%** of the electrons' **kinetic energy** is converted into **X-rays**. The rest is converted into **heat**, so, to avoid overheating, the tungsten anode is **rotated** at about 3000 rpm. It's also **mounted** on **copper** — this **conducts** the heat away effectively.

Beam Intensity is Power per Unit Area

The **intensity** of the X-ray beam is the **power (energy per second) per unit area** passing through a surface (at right angles). There are two ways to increase the **intensity** of the X-ray beam:

1) Increase the **tube voltage**. This gives the electrons **more kinetic energy**. Higher energy electrons can **knock out** electrons from shells **deeper** within the tungsten atoms.

2) Increase the **current** supplied to the filament. This liberates **more electrons per second**, which then produce **more X-ray photons per second**.

Charlotte thought landing on her head would get top marks for beam intensity.

Radiographers try to Produce a Sharp Image and Minimise the Radiation Dose

Medical X-rays are a compromise between producing really sharp, clear images, whilst keeping the amount of radiation the patient is exposed to as low as possible. To do this, radiographers:

1) Put the **detection plate close** to and the **X-ray tube far** from the patient and make sure the patient **keeps still**.

2) Put a **lead grid** between the patient and film to **stop** scattered radiation 'fogging' the film and **reducing contrast**.

3) Use an **intensifying screen** next to the film surface. This consists of crystals that **fluoresce** — they **absorb X-rays** and re-emit the energy as **visible light**, which helps to develop the photograph quickly. A shorter exposure time is needed, keeping the patient's radiation dose lower.

X-Ray Imaging

X-Rays are Attenuated when they Pass Through Matter

When X-rays pass through matter (e.g. a patient's body), they are **absorbed** and **scattered**. The intensity (**I**) of the X-ray beam **decreases** (attenuates) **exponentially** with **distance from the surface** (**x**), according to the material's attenuation coefficient (**μ**), as the equation shows.

$$I = I_0 e^{-\mu x}$$

1) **Half-value thickness**, $x_{\frac{1}{2}}$, is the thickness of material required to **reduce** the **intensity** to **half** its **original value**. This depends on the **attenuation coefficient** of the material, and is given by:

$$x_{\frac{1}{2}} = \frac{\ln 2}{\mu}$$

2) The **mass attenuation coefficient**, μ_m, for a material of density ρ is given by:

$$\mu_m = \frac{\mu}{\rho}$$

X-rays are Absorbed More by Bone than Soft Tissue

X-rays are **attenuated** by **absorption** and **scattering**. The **three** main **causes** of this are:

1) The **photoelectric effect** — a **photon** with around **30 keV** of energy is absorbed by an **electron**, which is **ejected** from its atom. The gap in the **electron shell** is filled by another **electron**, which emits a **photon**.

2) **Compton scattering** — a **photon** with around **0.5-5 MeV** of energy knocks an **electron** out of an **atom**, which causes the **photon** to **lose energy** and be **scattered**.

3) **Pair production** — a **high** (> 1.1 MeV) **energy** photon **decays** into an **electron** and a **positron**.

How much **energy is absorbed** by a **material** depends on its **atomic number** — so tissues containing atoms with **different atomic numbers** (e.g. **soft tissue** and **bone**) will **contrast** in the X-ray image.

If the tissues in the region of interest have similar attenuation coefficients then artificial **contrast media** can be used — e.g. **barium meal**. **Barium** has a **high atomic number**, so it shows up clearly in X-ray images and can be followed as it moves along the patient's digestive tract.

Fluoroscopy and CT Scans use X-rays

1) **Moving images** can be created by **X-ray fluoroscopy**, using a **fluorescent screen** and an **image intensifier**.

2) **Computerised axial tomography** (CT or CAT) scans produce an image of a **two-dimensional slice** through the body. An **X-ray beam rotates** around the body and is picked up by thousands of **detectors**. A computer works out how much attenuation has been caused by each part of the body and produces a very **high quality** image. However, the machines are **expensive**.

3) Both these techniques involve a **high radiation dose** for the patient.

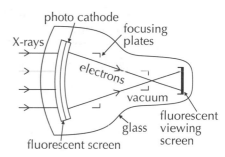

Practice Questions

Q1 Draw a diagram of an X-ray tube.

Q2 What measures can be taken to produce a high quality X-ray image while reducing the patient's radiation dose?

Exam Question

Q1 The half-value thickness for aluminium is 3 mm for 30 keV X-ray photons.

(a) Explain what is meant by the term 'half-value thickness'. [1 mark]

(b) What thickness of aluminium would reduce the intensity of a homogeneous beam of X-rays at 30 keV to 1% of its initial value? [4 marks]

I've got attenuation coefficient disorder — I get bored really easily...

X-ray images are just shadow pictures — bones absorb X-rays, stop them reaching the film and create a white 'shadow'.

Ultrasound Imaging

Ultrasound is a 'sound' with higher frequencies than we can hear.

Ultrasound has a Higher Frequency than Humans can Hear

1) Ultrasound waves are **longitudinal** waves with **higher frequencies** than humans can hear (>20 000 Hz).

2) For **medical** purposes, frequencies are usually from **1** to **15 MHz**.

3) When an ultrasound wave meets a **boundary** between two **different materials**, some of it is **reflected** and some of it passes through (undergoing **refraction** if the **angle of incidence** is **not 90°**).

4) The **reflected waves** are detected by the **ultrasound scanner** and are used to **generate an image**.

The Amount of Reflection depends on the Change in Acoustic Impedance

1) The **acoustic impedance**, **Z**, of a medium is defined as: $\boxed{Z = \rho c}$
   Z has units of $kg\,m^{-2}s^{-1}$.

2) Say an ultrasound wave travels through a material with an impedance Z_1. It hits the boundary between this material and another with an impedance Z_2. The incident wave has an intensity of I_i.

3) If the two materials have a **large difference** in **impedance**, then **most** of the energy is **reflected** (the intensity of the reflected wave I_r will be high). If the impedance of the two materials is the **same** then there is **no reflection**.

4) The **fraction** of wave **intensity** that is reflected is called the **intensity reflection coefficient**, α.

$$\alpha = \frac{I_r}{I_i} = \left(\frac{Z_2 - Z_1}{Z_2 + Z_1}\right)^2$$

You don't need to learn this equation. Just practise using it.

There are Advantages and Disadvantages to Ultrasound Imaging

ADVANTAGES:

1) There are **no** known **hazards** — in particular, **no** exposure to **ionising radiation**.

2) It's good for imaging **soft tissues**, since you can obtain **real-time** images — X-ray fluoroscopy can achieve this, but involves a huge dose of radiation.

3) Ultrasound devices are relatively **cheap** and **portable**.

DISADVANTAGES:

1) Ultrasound **doesn't penetrate bone** — so it **can't** be used to **detect fractures** or examine the **brain**.

2) Ultrasound **cannot** pass through **air spaces** in the body (due to the **mismatch** in **impedance**) — so it can't produce images from behind the lungs.

3) The **resolution** is **poor** (about 10 times worse than X-rays), so you **can't see** fine **detail**.

Ultrasound Images are Produced Using the Piezoelectric Effect

1) **Piezoelectric crystals** produce a **potential difference** when they are **deformed** (squashed or stretched) — the rearrangement in structure displaces the **centres of symmetry** of their electric **charges**.

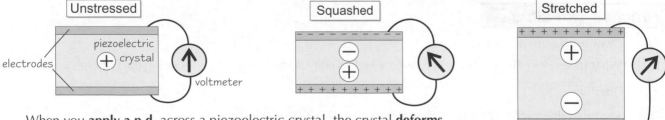

2) When you **apply a p.d.** across a piezoelectric crystal, the crystal **deforms**. If the p.d. is **alternating**, then the crystal **vibrates** at the **same frequency**.

3) A piezoelectric crystal can act as a **receiver** of **ultrasound**, converting **sound waves** into **alternating voltages**, and also as a **transmitter**, converting **alternating voltages** into **sound waves**.

4) Ultrasound devices use **lead zirconate titanate** (**PZT**) crystals. The **thickness** of the crystal is **half the wavelength** of the ultrasound that it produces. Ultrasound of this frequency will make the crystal **resonate** (like air in an open pipe — see p.98) and produce a large signal.

5) The PZT crystal is **heavily damped**, to produce **short pulses** and **increase** the **resolution** of the device.

Ultrasound Imaging

You need a **Coupling Medium** between the **Transducer** and the **Body**

1) **Soft tissue** has a very different **acoustic impedance** from **air**, so almost all the ultrasound **energy** is **reflected** from the surface of the body if there is air between the **transducer** and the **body**.

2) To avoid this, you need a **coupling medium** between the transducer and the body — this **displaces** the **air** and has an impedance much closer to that of body tissue. **Coupling media** are an example of **impedance matching**.

3) The coupling medium is usually an **oil** or **gel** that is smeared onto the skin.

The **A-Scan** is a **Range Measuring** System

1) The **amplitude scan** (**A-Scan**) sends a short **pulse** of ultrasound into the body simultaneously with an **electron beam** sweeping across a cathode ray oscilloscope (**CRO**) screen.

2) The scanner receives **reflected** ultrasound pulses that appear as **vertical deflections** on the CRO screen.

3) **Weaker** pulses (that have travelled further in the body and **arrive later**) are **amplified** more to avoid the loss of valuable data — this process is called **time-gain compensation** (**TGC**).

4) The **horizontal positions** of the reflected pulses indicate the **time** the 'echo' took to return, and are used to work out **distances** between structures in the body (e.g. the **diameter** of a **baby's head** in the uterus).

5) A **stream** of pulses can produce a **steady image** on the screen due to **persistence of vision**, although modern CROs can store a digital image after just one exposure.

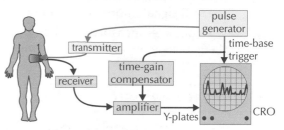

In a **B-Scan**, the **Brightness** Varies

1) In a **brightness scan** (**B-Scan**), the electron beam sweeps **down** the screen rather than across.

2) The amplitude of the reflected pulses is displayed as the **brightness** of the spot.

3) You can use a **linear array** of transducers to produce a **two-dimensional** image.

> **Ultrasound imaging** can also be used to measure the **speed of blood flow**.
> If the **ultrasound waves** reflect off something that is **moving** (e.g. **blood**), their **frequency** will be **shifted** according to the **Doppler effect** (see p.156). **How much** the reflected waves are **shifted** depends on **how fast** the blood is moving — so by **measuring the shift**, you can work out the **speed of blood flow**.

Practice Questions

Q1 What are the main advantages and disadvantages of imaging using ultrasound?

Q2 How are ultrasound waves produced and received in an ultrasound transducer?

Q3 Define acoustic impedance.

Exam Questions

Q1 (a) What fraction of intensity is reflected when ultrasound waves pass from air to soft tissue?
Use $Z_{air} = 0.430 \times 10^3$ kgm^{-2}s^{-1}, $Z_{tissue} = 1630 \times 10^3$ kgm^{-2}s^{-1}. [2 marks]

(b) Calculate the ratio between the intensity of the ultrasound that **enters** the body when a coupling gel is used ($Z_{gel} = 1500 \times 10^3$ kgm^{-2}s^{-1}) and when none is used. Give your answer to the nearest power of ten. [4 marks]

Q2 (a) The acoustic impedance of a certain soft tissue is 1.63×10^6 kgm^{-2}s^{-1} and its density is 1.09×10^3 kgm^{-3}. Show that ultrasound travels with a velocity of 1.50 kms^{-1} in this medium. [2 marks]

(b) The time base on a CRO was set to be 50 μscm^{-1}. Reflected pulses from either side of a fetal head are 2.4 cm apart on the screen. Calculate the diameter of the fetal head if the ultrasound travels at 1.5 kms^{-1}. [4 marks]

Ultrasound — Mancunian for 'très bien'

You can use ultrasound to make images in cases where X-rays would do too much damage — like to check up on the development of a baby in the womb. You have to know what you're looking for though, or it just looks like a blob.

Magnetic Resonance Imaging

Magnetic Resonance Imaging, or MRI to you and me, is yet another form of non-invasive diagnostic imaging — enjoy.

Magnetic Resonance can be used to Create Images

1) The patient lies in the centre of a huge **superconducting magnet** that produces a **uniform magnetic field**. The magnet needs to be **cooled** by **liquid helium** — this is partly why the scanner is so **expensive**.

2) Radio frequency **coils** are used to transmit **radio waves**, which **excite hydrogen nuclei** in the patient's body.

3) When the radio waves are switched off, the hydrogen nuclei relax and emit electromagnetic energy — this is the **MRI signal** (more details below). The radio frequency coils **receive the signal** and send it to a **computer**.

4) The computer **measures** various quantities of the MRI signal — amplitude, frequency, phase — and **analyses** them to generate an **image** of a **cross-section** through the body.

Contrast can be Controlled by Varying the Pulses of Radio Waves

1) Radio waves are applied in **pulses**. Each short pulse **excites** the hydrogen nuclei and then allows them to **relax** and emit a signal. The response of **different tissue types** can be enhanced by varying the **time between pulses**.

2) Tissues consisting of **large molecules** such as **fat** are best imaged using **rapidly repeated pulses**. This technique is used to image the internal **structure** of the body.

3) Allowing **more time** between pulses enhances the response of **watery** substances. This is used for **diseased** areas.

MRI has Advantages and Disadvantages

ADVANTAGES:

1) There are **no** known **side effects**.

2) An image can be made for any slice in any **orientation** of the body.

3) High quality images can be obtained for **soft tissue** such as the **brain**.

4) **Contrast** can be **weighted** in order to investigate different situations.

DISADVANTAGES:

1) The imaging of **bones** is very **poor**.

2) Some people suffer from **claustrophobia** in the scanner.

3) Scans can be **noisy** (due to the switching of the gradient magnets, see page 149) and take a **long time**.

4) MRI can't be used on people with **pacemakers** or some **metal implants** — the strong magnetic fields would be very harmful.

5) Scanners **cost millions** of pounds.

Atomic Nuclei can Behave like Magnets

1) **Protons** and **neutrons** possess a quantum property called **spin**, which makes them behave like **tiny magnets**.

2) If a nucleus has **even numbers** of **protons** and **neutrons** then the magnetic effects **cancel out**. A nucleus with an **odd number** of **protons** or **neutrons** has a **net spin** and is slightly **magnetic**.

3) The most important nucleus for **magnetic resonance imaging (MRI)** is **hydrogen**, which human bodies contain a lot of. A **hydrogen nucleus** has just **one proton**.

Protons Align themselves in a Magnetic Field

1) Normally, protons are orientated **randomly**, so their magnetic fields cancel out. When a strong **external magnetic field** is applied, as in an MRI scanner, the protons **align** themselves with the **magnetic field lines**.

2) Protons in **parallel alignment** point in the **same direction** as the external **magnetic field**. **Antiparallel alignment** means the protons point in the **opposite direction to the field**.

3) Protons align in these two directions in almost equal numbers (about 7 more per million are parallel). The **different alignments** correspond to **different energy levels**. The nuclei can **flip** between the two by **emitting** or **absorbing** a specific amount of **energy**.

$$\Delta E \approx 2 \ \mu eV$$

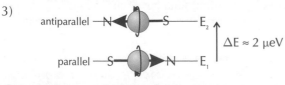

Magnetic Resonance Imaging

Protons in a Magnetic Field Precess at the Larmor Frequency

1) Nuclei in a magnetic field don't just stay still — they **precess** (wobble) **around** the **magnetic field lines** (like a **spinning top** precessing around gravitational field lines).

2) They **don't** all precess **in phase** with each other.

3) All the protons **precess** at the **same frequency** — the **Larmor frequency**, *f*. The value of *f* depends on the **strength** of the **magnetic field**, B_0.

$$f = \frac{\gamma B_0}{2\pi}$$

γ = gyromagnetic ratio (in HzT^{-1}),
B_0 = magnetic flux density of the external field (in T).

4) For protons, $\dfrac{\gamma}{2\pi} = 42.57$ $MHzT^{-1}$.

In an MRI scanner, the magnetic field, B_0 is about **1-2 teslas**, which puts *f* in the **radio** frequency range.

There are also three **gradient magnets** — electromagnets which produce a very small field (superimposed on the main one) that gradually **varies** from place to place. The **Larmor frequency** depends on the **total magnetic flux density**, so **different frequencies** of radio wave can be used to target **different sites** within the body.

Radio Waves can Make Protons Resonate

1) Protons have a **natural frequency** of oscillation in a magnetic field which is **equal** to the **Larmor frequency**, f. **Radio waves** at this frequency can make protons **resonate**.

2) The protons **absorb energy** from the radio waves and **flip** from **parallel** to **antiparallel alignment**.

3) Radio waves at the Larmor frequency also make the protons precess **in phase** with each other, producing a **rotating magnetic field** at **right angles** to the external field.

4) When the **external radio waves stop**, the protons return to their **original states** and **emit electromagnetic energy** as they do so — this emitted energy is the MRI signal.

5) The time taken for the protons to return to their original state is called the **relaxation time**, and is about **one second**.

6) **Relaxation times** depend on what molecules surround the protons, so they **vary** for **different tissue types**.

Practice Questions

Q1 Define the Larmor frequency and explain how radio waves at this frequency can cause resonance.

Q2 What is 'relaxation time' and how is this used to generate contrast within the images?

Q3 How do MRI scanners target protons from particular parts of the body?

Q4 Give two advantages of MRI compared to X-ray imaging.

Exam Questions

Q1 (a) Using a diagram, explain what is meant by the term 'precession'. [2 marks]

(b) What is the 'Larmor frequency'? Suggest a typical value for the Larmor frequency in an MRI scanner. [2 marks]

Q2 Outline how an MRI scanner is used to produce an image of a section of a patient's body. [5 marks]

Q3 Discuss the advantages and disadvantages of MRI scanning as a medical imaging technique. [6 marks]

Precession — protons on parade...

OK, so it hasn't been the easiest of pages. But at least now you know why people sit in vats of baked beans to raise money for their local hospital to buy an MRI scanner. Though perhaps you need A2 Psychology to understand the beans part.

Medical Uses of Nuclear Radiation

Radiation can be incredibly useful in medicine, but any use of radiation carries some risk.

Medical Tracers are Used to Diagnose the Function of Organs

Medical tracers are **radioactive substances** that are used to show tissue or **organ function**.
Other types of imaging, **e.g. X-rays**, only show the **structure** of organs — medical tracers show **structure and function**.

Medical tracers usually consist of a **radioactive isotope** — e.g. **technetium-99m** — bound to a **substance** that is **used** by the **body** — e.g. **glucose** or **water**. The tracer is **injected** into or **swallowed** by the patient and then **moves** through the **body** to the region of interest. **Where** the tracer goes depends on the **substance** the isotope is bound to — i.e. it goes anywhere that the substance would **normally go**, and is used how that substance is **normally used**. The **radiation emitted** is **recorded** (e.g. by a **gamma camera** or **PET scanner**, see below) and an **image** of inside the patient produced.

1) Tracers can show areas of **damaged tissue** in the heart by detecting areas of **decreased blood flow**. This can reveal **coronary artery disease** and damaged or dead heart muscle caused by **heart attacks**.

2) They can identify **active cancer tumours** by showing **metabolic activity** in tissue. **Cancer cells** have a **much higher metabolism** than healthy cells because they're growing fast, so take up more tracer.

3) Tracers can show **blood flow and activity** in the **brain**. This helps **research** and **treat** neurological conditions like Parkinson's, Alzheimer's, epilepsy, depression, etc.

Technetium-99m is widely used in medical tracers because it emits **γ-radiation**, has a **half-life of 6 hours** (long enough for data to be recorded, but short enough to limit the radiation to an acceptable level) and **decays** to a **much more stable isotope**.

Gamma Cameras Detect Gamma Radiation

The **γ-rays** emitted by **radiotracers** injected into a patient's body are detected using a **gamma camera**.
Gamma cameras (like the one shown **below**) consist of **five** main parts:

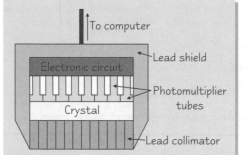

1) **Lead shield** — **stops radiation** from **other sources** entering the camera.

2) **Lead collimator** — a **piece of lead** with thousands of **vertical holes** in it — only γ-rays **parallel** to the holes can **pass through**.

3) **Sodium iodide crystal** — emits a **flash of light** (**scintillates**) whenever a **γ-ray** hits it.

4) **Photomultiplier tubes** — **detect** the flashes of **light** from the crystal and turn them into **pulses of electricity**.

5) **Electronic circuit** — **collects** the **signals** from the photomultiplier tubes and sends them to a **computer** for processing into an **image**.

PET Scanning Involves Positron/Electron Annihilation

1) The patient is injected with a substance used by the body, e.g. glucose, containing a **positron-emitting** radiotracer with a **short half-life**, e.g. ^{13}N, ^{15}O, ^{18}F.

2) The patient is left for a time to allow the radiotracer to **move through the body** to the organs.

3) **Positrons** emitted by the radioisotope collide with **electrons** in the organs, causing them to **annihilate**, emitting **high-energy gamma rays** in the process.

4) **Detectors** around the body record these **gamma rays**, and a computer builds up a **map of the radioactivity** in the body.

5) The **distribution of radioactivity** matches up with **metabolic activity**. This is because **more** of the radioactive glucose (or whatever) injected into the patient is taken up and **used** by cells that are **doing more work** (cells with an **increased metabolism**, in other words).

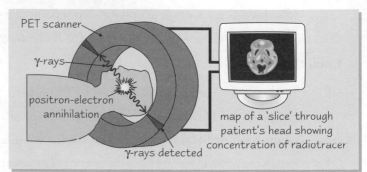

map of a 'slice' through patient's head showing concentration of radiotracer

Medical Uses of Nuclear Radiation

Ionising Radiation is Used When the Benefits Outweigh the Risks

X-rays, γ-rays, and α and β particles are all classed as **ionising radiation**. When they **interact** with matter they **ionise atoms** or **molecules** to form **ions** — usually by **removing an electron** — and this can **damage** cells. Cell damage is bad news — it can cause:

1) **Cell mutations** and **cancerous tumours** by altering or damaging the cell's DNA.
2) **Cell sterility** by stopping the cell from reproducing.
3) **Cell death** — the cell is destroyed completely.

The **macroscopic effects** of ionising radiation (i.e. the large-scale effects) include **tumours**, **skin burns**, **sterility**, **radiation sickness**, **hair loss** and **death** — nice. The result is that radiation is only used when the **benefits** to the patient **outweigh** the **risks** — i.e. **radiation** doses are **limited** and only used when it's **absolutely necessary**.

Exposure, Absorbed Dose and Effective Dose

1) **Exposure** is defined as the **total charge produced** by ionising radiation **per unit mass of air**. The unit of exposure is the C kg⁻¹.

> Air is used since its average atomic number is similar to that of body tissue.

2) The **absorbed dose** is more useful in medical physics because it is a measure of the **energy absorbed per unit mass**. The unit is the **gray (Gy)** where 1 Gy = 1 Jkg⁻¹.

3) The **effective dose**, **H**, takes into account the fact that the **amount of damage** to body tissue depends on the **type of ionising radiation**. The unit is the **sievert (Sv)** where 1 Sv = 1 Jkg⁻¹.

$$H = Q \times D$$

where H is the effective dose (Sv), Q is the quality factor, and D is the absorbed dose (Gy)

Radiation	Q
X-ray, β, γ	1
neutrons	5 - 20
α	20

4) **Each type** of radiation is assigned a weighted **quality factor**, **Q** — the **greater** the **damage** produced by a type of radiation, the **higher** the quality factor.

> **Example** Which absorbed dose would cause the greatest damage to a cell, 5 Gy of α or 10 Gy of β radiation?
> To compare the potential damage to the cell you need to look at the effective dose of each type of radiation.
> $H_\alpha = 20 \times 5 = 100$ Sv $H_\beta = 10 \times 1 = 10$ Sv
> So **5 Gy of α** radiation would cause **more** cell damage than **10 Gy of β**.

Practice Questions

Q1   Why are medical tracers useful? Why are they dangerous?
Q2   What are the five main parts of a gamma camera?
Q3   Describe how an image is formed in PET scanning.
Q4   What's the difference between absorbed dose and effective dose?

Exam Questions

Q1   A doctor suspects that his patient has a cancerous tumour.
Describe a non-invasive technique that could be used to confirm the doctor's diagnosis.   [3 marks]

Q2   A man of mass 70 kg accidentally swallows a source of α radiation (with a radiation quality factor of 20).
The source emits 3×10^{10} particles per second (constant over this time period) and the energy of each particle is 8×10^{-13} J. The source is removed after 1000 seconds.
(a) Assuming that all the ionising energy is distributed uniformly around the man's body,
what is the absorbed dose that he receives?   [2 marks]
(b) Calculate the effective dose.   [1 mark]

The biological effects of a page on radiation — a sore head...

Hoo-bleeding-rah — at last, you've made it to the end of a proper beast of a section. But before you tootle off, just make sure you know the difference between absorbed dose and effective dose and can work each one out — there's bound to be a question on them in your exam. Oh, and don't forget to come back for the next section — it's a real good 'un.

The Solar System & Astronomical Distances

The Meaning of Life, Part 6: A2 Physics...

Our **Solar System** Contains the **Sun**, **Planets**, **Satellites**, **Asteroids** and **Comets**

1) The **Universe** is **everything** that exists — this includes plenty you can see, like **stars** and **galaxies**, and plenty that you can't see, like **microwave radiation** (see page 157).

2) **Galaxies**, like our very own **Milky Way galaxy**, are clusters of **stars** and **planets** that are held together by gravity.

3) Inside the Milky Way is our **Solar System**, which consists of the **Sun** and all of the objects that **orbit** it:

Asteroid belt

The planets (in order): **Mercury, Venus, Earth, Mars, Jupiter, Saturn, Uranus** and **Neptune** (as well as the asteroid belt) all have nearly **circular** orbits. We used to call Pluto a planet too, but it's been reclassified now.

Remember — planets, moons and comets don't emit light; they just reflect it.

4) The orbits of the **comets** we see are **highly elliptical**. Comets are "**dirty snowballs**" that we think usually orbit the Sun about **1000 times further away** than **Pluto** does (in the "Oort cloud"). Occasionally one gets **dislodged** and heads towards the Sun. It follows a new elliptical orbit, which can take **millions of years** to complete. Some comets (from closer in than the Oort cloud) follow a **smaller orbit** and they return to swing round the Sun more regularly. The most famous is **Halley's comet**, which orbits in **76 years**.

Distances and **Velocities** in the Solar System can be Measured using **Radar**

1) **Radio telescopes** can be used to send **short pulses** of **radio waves** towards a planet or asteroid (a rock flying about the Solar System), which **reflect** off the surface and bounce back.

2) The telescope picks up the reflected radio waves, and the **time taken** (*t*) for them to return is measured.

3) Since we know the **speed** of radio waves (**speed of light, c**) we can work out the **distance, d**, to the object using:

$$2d = ct$$

It's 2d, not just d, because the pulse travels twice the distance to the object — there and back again.

4) If **two** short pulses are sent a certain **time interval** apart, you can measure the **distance** an object has moved in that time. From this time and distance, you can calculate the **average speed** of the object **relative** to Earth. More accurate measurements can be made using Doppler shifts (see p. 156).

Distances in the Solar System are Often Measured in **Astronomical Units (AU)**

1) From **Copernicus** onwards, astronomers were able to work out the **distance** the **planets** are from the Sun **relative** to the Earth, using **astronomical units** (AU). But they could not work out the **actual distances**.

> One **astronomical unit** (AU) is defined as the **mean distance** between the **Earth** and the **Sun**.

2) The **size** of the AU wasn't known accurately until 1769 — when it was carefully **measured** during a **transit of Venus** (when Venus passed between the Earth and the Sun).

Another Measure of Distance is the **Light-Year (ly)**

1) All **electromagnetic waves** travel at the **speed of light**, *c*, in a vacuum ($c = 3.00 \times 10^8$ ms^{-1}).

> The **distance** that electromagnetic waves travel through a vacuum in **one year** is called a **light-year** (ly).

2) If we see the light from a star that is, say, **10 light-years away** then we are actually seeing it as it was **10 years ago**. The further away the object is, the further **back in time** we are actually seeing it.

3) **1 ly** is equivalent to about **63 000 AU**.

The Solar System & Astronomical Distances

The **Distance** to **Nearby Stars** can be Measured in **Parsecs**

1) Imagine you're in a **moving car**. You see that (stationary) objects in the **foreground** seem to be **moving faster** than objects in the **distance**. This **apparent change in position** is called **parallax**.

2) Parallax is measured in terms of the **angle of parallax**. The **greater** the **angle**, the **nearer** the object is to you.

3) The distance to **nearby stars** can be calculated by observing how they **move relative** to **very distant stars** when the Earth is in **different parts** of its **orbit**. This gives a **unit** of distance called a **parsec (pc)**.

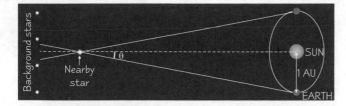

A star is exactly **one parsec (pc)** away from Earth if the **angle of parallax**,

$$\theta = 1 \text{ arcsecond} = \left(\frac{1}{3600}\right)^{\circ}$$

Important **Sizes** and **Conversions**

You need to be able to state the length of a light-year and a parsec in metres. So here's a handy table to help you.

Unit of Distance	Light-year (ly)	Parsec (pc)	Astronomical Unit (AU)
Approximate Length in metres	9.46×10^{15}	3.09×10^{16}	1.50×10^{11}

You don't need to learn the length of an AU — it's just here for comparison.

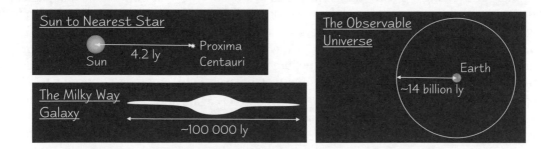

When we look at the stars we're looking **back in time**, and we can only see as far back as the **beginning of the Universe**. So the **size** of the **observable Universe** is the **age** of the Universe multiplied by the **speed of light**.

Practice Questions

Q1 What are the principal contents of : a) the Universe, b) our Solar System?

Q2 What is meant by: a) an astronomical unit, b) a parsec, and c) a light-year?

Q3 How do we measure the distance to objects in the Solar System using radar?

Exam Questions

Q1 Outline the main differences between planets and comets. [5 marks]

Q2 (a) Give the definition of a *light-year*. [1 mark]

(b) Calculate the distance of a light-year in metres. [2 marks]

(c) Why is the size of the observable universe limited by the speed of light? [2 marks]

So — using a ruler's out of the question then...

Don't bother trying to get your head round these distances — they're just too big to imagine. Just learn the powers of ten and you'll be fine. Make sure you understand the definition of a parsec — it's a bit of a weird one.

Stellar Evolution

Stars go through several different stages in their lives — from clouds of dust and gas, to red giants to white dwarfs...

Stars Begin as Clouds of Dust and Gas

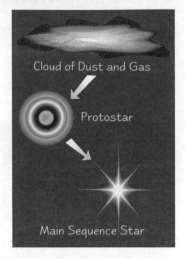

Cloud of Dust and Gas

Protostar

Main Sequence Star

1) Stars are born in a **cloud** of **dust** and **gas**, most of which was left when previous stars blew themselves apart in **supernovae**. The denser clumps of the cloud **contract** (very slowly) under the force of **gravity**.

2) When these clumps get dense enough, the cloud fragments into regions called **protostars**, that continue to contract and **heat up**.

3) Eventually the **temperature** at the centre of a protostar reaches a **few million degrees**, and **hydrogen nuclei** start to **fuse** together to form helium (see page 143 for more on nuclear fusion).

4) This releases an **enormous** amount of **energy** and creates enough **pressure** (radiation pressure) to stop the **gravitational collapse**.

5) The star has now reached the **MAIN SEQUENCE** and will stay there, relatively **unchanged**, while it fuses hydrogen into helium.

Main Sequence Stars become Red Giants when they Run Out of Fuel

1) Stars spend most of their lives as **main sequence** stars. The **pressure** produced from **hydrogen fusion** in their **core balances** the **gravitational force** trying to compress them. This stage is called **core hydrogen burning**.

2) When the **hydrogen** in the **core** runs out nuclear fusion **stops**, and with it the **outward pressure stops**. The core **contracts** and **heats up** under the **weight** of the star.

3) The material **surrounding** the core still has **plenty of hydrogen**. The **heat** from the contracting **core** raises the **temperature** of this material enough for the hydrogen to **fuse**. This is called **shell hydrogen burning**. (Very low-mass stars stop at this point. They use up their fuel and slowly fade away...)

4) The core continues to contract until, eventually, it gets **hot** enough and **dense** enough for **helium** to **fuse** into **carbon** and **oxygen**. This is called **core helium burning**. This releases a **huge** amount of energy, which **pushes** the **outer layers** of the star outwards. These outer layers **cool**, and the star becomes a **RED GIANT**.

5) When the **helium** runs out, the carbon-oxygen core **contracts again** and heats a **shell** around it so that helium can fuse in this region — **shell helium burning**.

Low Mass Stars (like the Sun) Eject their Shells, leaving behind a White Dwarf

1) In low-mass stars, the **carbon-oxygen core isn't hot enough** for any further **fusion** and so it continues to **contract** under its own **weight**. Once the core has shrunk to about **Earth-size**, **electrons** exert enough pressure (**electron degeneracy pressure**) to stop it collapsing any more (fret not — you don't have to know how).

2) The **helium shell** becomes more and more **unstable** as the core contracts. The star **pulsates** and **ejects** its outer layers into space as a **planetary nebula**, leaving behind the dense core.

3) The star is now a very **hot**, **dense solid** called a **WHITE DWARF**, which will simply **cool down** and **fade away**.

The Sun is a Main Sequence Star

The **Sun** might seem quite **special** to us on Earth — it's the reason we're all here after all. But really it's just like any other **low mass star** — it started off as a **cloud of dust and gas** and evolved to be the **main sequence star** we see today. Of course, this means it will most likely become a **red giant** and then finally fizzle out as a **white dwarf** — sob.

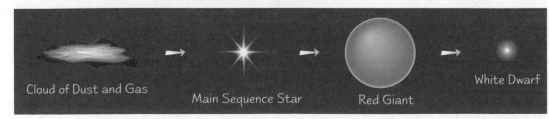

Cloud of Dust and Gas   Main Sequence Star   Red Giant   White Dwarf

Stellar Evolution

High Mass Stars have a Shorter Life and a more Exciting Death

The **mass** of a star will determine how it **evolves**. **Low to medium mass** stars (like the **Sun**) follow the sequence on the previous page. **High mass stars** (i.e. with around **8 times** the mass of the Sun, or more) follow the **sequence below**.

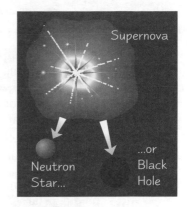

1) Stars with a **large mass** have a **lot of fuel**, but they use it up **more quickly** and don't spend so long as main sequence stars.

2) When they are **red giants** the **'core burning to shell burning'** process can continue beyond the fusion of helium, building up layers in an **onion-like structure** to become **SUPER RED GIANTS**.

3) For **really massive** stars, fusion can go all the way up to **iron**. Nuclear fusion **beyond iron** isn't **energetically favourable**, though, so once an iron core is formed then very quickly it's goodbye star.

4) The star explodes cataclysmically in a **SUPERNOVA**, leaving behind a **NEUTRON STAR** or (if the star was massive enough) a **BLACK HOLE**.

Massive Stars go out with a Bit of a Bang

1) When the core of a star runs out of fuel, it starts to **contract** — forming a white dwarf core.

2) If the star is **massive enough**, though, **electron degeneracy** can't stop the core contracting. This happens when the mass of the core is more than **1.4 times** the mass of the **Sun**.

3) The electrons get **squashed** onto the atomic **nuclei**, combining with protons to form **neutrons** and **neutrinos**.

4) The core suddenly collapses to become a **NEUTRON STAR**, which the outer layers then **fall** onto.

5) When the outer layers **hit** the surface of the **neutron star** they **rebound**, setting up huge **shockwaves**, ripping the star apart and causing a **supernova**. The light from a supernova can briefly outshine an **entire galaxy**!

Neutron Stars are Very Dense — But not as Dense as Black Holes

Neutron stars are incredibly **dense** (about 4×10^{17} kgm^{-3}).

They're **very small**, typically about 20 km across, and they can **rotate very fast** (up to 600 times a second).

They emit **radio waves** in two beams as they rotate. These beams sometimes sweep past the Earth and can be observed as **radio pulses** rather like the flashes of a lighthouse. These rotating neutron stars are called **PULSARS**.

If the **core** of a star is more than **3 times** the **Sun's mass**, it can't withstand the gravitational forces and collapses to form a **black hole**. The physics of black holes is mind-boggling, and luckily you don't need to know it.

Practice Questions

Q1 How are stars formed?

Q2 Outline the life cycle of the Sun, including its probable evolution and death.

Q3 What causes a star to evolve from the main sequence to become a red giant?

Q4 Describe a white dwarf and a neutron star. What are the main differences between them?

Exam Question

Q1 Outline the main differences between the evolution of high mass and low mass stars, starting from when they first become main sequence stars. [6 marks]

Live fast — die young...

The more massive a star, the more spectacular its life cycle. The most massive stars burn up the hydrogen in their core so quickly that they only live for a fraction of the Sun's lifetime — but when they go, they do it in style.

The Big Bang Model of the Universe

Everyone's heard of the Big Bang theory — well here's some evidence for it.

An **Infinite Universe** leads to **Olbers' Paradox**

It's easy to imagine that the Earth is at the **centre of the Universe**, or that there's something really **special** about it. **Earth** is special to us because we **live here** — but on a **universal scale**, it's just like any other lump of rock.

1) The **demotion** of **Earth** from anything special is taken to its logical conclusion with the **cosmological principle**...

> **COSMOLOGICAL PRINCIPLE:** on a **large scale** the Universe is **homogeneous** (every part is the same as every other part) and **isotropic** (everything looks the same in every direction) — so it doesn't have a **centre**.

2) Until the **1930s**, cosmologists believed that the Universe was **infinite** in both **space** and **time** (that is, it had always existed) and **static**. This seemed the **only way** it could be **stable** using **Newton's law** of gravitation. Even **Einstein modified** his theory of **general relativity** to make it consistent with the **Steady-State Universe**.

3) In the 1820s, though, an astronomer called **Olbers** noticed a **big problem** with this model of the Universe.

> If stars (or galaxies) are **spread randomly** throughout an **infinite** Universe then **every possible line of sight** must contain a **star**. Calculations show that this should make the **whole** night sky **uniformly bright**.

This problem is called **Olbers' paradox** and clearly there's a **contradiction** with an infinite and static Universe.

The **Doppler Effect** — the **Motion** of a Wave's **Source** Affects its **Wavelength**

1) Imagine an ambulance driving past you. As it moves **towards you** its siren sounds **higher-pitched**, but as it **moves away**, its **pitch** is **lower**. This change in **frequency** and **wavelength** is called the **Doppler shift**.

2) The frequency and the wavelength **change** because the waves **bunch together** in **front** of the source and **stretch out behind** it. The **amount** of stretching or bunching together depends on the **velocity** of the **source**.

3) This happens with light too — when a **light source** moves **away** from us, the wavelengths become **longer** and the frequencies become lower. This shifts the light towards the **red** end of the spectrum and is called **redshift**.

4) When a light source moves **towards** us, the **opposite** happens and the light undergoes **blueshift**.

5) The amount of redshift or blueshift is determined by the following formula:

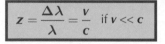

z is the redshift, $\Delta\lambda$ is the difference between the observed wavelength and the emitted wavelength, λ is the emitted wavelength, v is the velocity of the source in the direction of the observer and c is the speed of light. ($v \ll c$ means "v is much less than c".)

Hubble Realised that the **Universe** is **Expanding**

1) The **spectra** from **galaxies** (apart from a few very close ones) all show **redshift** — so they're all **moving away** from us. The amount of **redshift** gives the **recessional velocity** — how fast the galaxy is moving away.

2) Plotting **recessional velocity** against **distance** shows that they're **proportional** — i.e. the **speed** that **galaxies move away** from us depends on **how far** away they are.

3) This suggests that the Universe is **expanding**, and gives rise to **Hubble's law**:

$$v = H_0 d$$

where v = recessional velocity in **kms⁻¹**, d = distance in **Mpc** and H_0 = **Hubble's constant** in **kms⁻¹Mpc⁻¹**.

4) Since distance is very difficult to measure, astronomers disagree on the value of H_0. It's generally accepted that H_0 lies somewhere between 50 kms⁻¹Mpc⁻¹ and 100 kms⁻¹Mpc⁻¹.

5) The **SI unit** for H_0 is s⁻¹. To get H_0 in SI units, you need v in ms⁻¹ and d in m (1 Mpc = 3.09×10^{22} m).

The **Expanding Universe** gives rise to the **Hot Big Bang Model**

The Universe is **expanding** and **cooling down**. So further back in time it must have been **smaller** and **hotter**. If you trace time back **far enough**, you get a **Hot Big Bang**.

> **THE HOT BIG BANG THEORY (HBB):**
> the Universe started off **very hot** and **very dense** (perhaps as an **infinitely hot, infinitely dense** singularity) and has been **expanding** ever since.

If the Universe began at a specific point in time, i.e. in the HBB, then it has a finite age.

The Big Bang Model of the Universe

The *Age* and *Observable Size* of the *Universe* Depend on H_0

1) If the Universe has been **expanding** at the **same rate** for its whole life, the **age** of the Universe is: $\boxed{t = 1/H_0}$
This is only an estimate since the Universe probably hasn't always been expanding at the same rate.

2) Unfortunately, since no one knows the **exact value** of H_0 we can only guess the Universe's age.
If $H_0 = 75 \text{ kms}^{-1}\text{Mpc}^{-1}$, then the age of the Universe $\approx 1/(2.4 \times 10^{-18} \text{ s}^{-1}) = 4.1 \times 10^{17}$ s = **13 billion years**.

3) The **absolute size** of the Universe is **unknown** but there is a limit on the size of the **observable Universe**. This is simply a **sphere** (with the Earth at its centre) with a **radius** equal to the **maximum distance** that **light** can travel during its **age**. So if $H_0 = 75 \text{ kms}^{-1}\text{Mpc}^{-1}$ then this sphere will have a radius of **13 billion light years**.

4) This gives a very simple solution to **Olbers' paradox**. If the observable Universe is **finite**, then there is **absolutely no reason** why every line of sight should include a star. Actually, most of them don't.

Cosmic Microwave Background Radiation — *More Evidence for the HBB*

1) The Hot Big Bang model predicts that loads of **electromagnetic radiation** was produced in the **very early Universe**. This radiation should **still** be observed today (it hasn't had anywhere else to go).

2) Because the Universe has **expanded**, the wavelengths of this cosmic background radiation have been **stretched** and are now in the **microwave** region.

3) Cosmic microwave background radiation (CMBR) was picked up **accidentally** by Penzias and Wilson in the 1960s.

4) In the late 1980s a satellite called the **Cosmic Background Explorer** (**COBE**) was sent up to have a **detailed look** at the radiation. It found a **continuous spectrum** corresponding to a **temperature** of about **3 K**.

5) The radiation is largely **isotropic** and **homogeneous**, which confirms the cosmological principle (see page 146).

6) There are **very tiny fluctuations** in temperature, which were at the limit of COBE's detection. These are due to tiny energy-density variations in the early Universe, and are needed for the initial 'seeding' of galaxy formation.

7) The background radiation also shows a **Doppler shift**, indicating the Earth's motion through space. It turns out that the **Milky Way** is rushing towards an unknown mass (the **Great Attractor**) at over a **million miles an hour**.

Practice Questions

Q1 State Olbers' paradox. How does the Hot Big Bang model resolve it?

Q2 What is Hubble's law? How can it be used to find the age of the Universe?

Q3 What is the cosmic background radiation?

Exam Questions

Q1 (a) State Hubble's law, explaining the meanings of all the symbols. [2 marks]

(b) What does Hubble's law suggest about the nature of the Universe? [2 marks]

(c) Assume $H_0 = +50 \text{ kms}^{-1}\text{Mpc}^{-1}$ (1 Mpc = 3.09×10^{22} m).

i) Calculate H_0 in SI units. [2 marks]

ii) Calculate an estimate of the age of the Universe, and hence the size of the observable Universe. [3 marks]

Q2 (a) A certain object has a redshift of 0.37. Estimate the speed at which it is moving away from us. [2 marks]

(b) Use Hubble's law to estimate the distance (in light years) that the object is from us.
(Take $H_0 = 2.4 \times 10^{-18} \text{ s}^{-1}$, 1 ly = 9.5×10^{15} m.) [2 marks]

Q3 Describe the main features of the cosmic background radiation and explain why its discovery was considered strong evidence for the Hot Big Bang model of the Universe.
In your answer, you should make clear how your explanation links with the evidence. [7 marks]

My Physics teacher was a Great Attractor — everyone fell for him...

The simple Big Bang model doesn't actually work — not quite, anyway. There are loads of little things that don't quite add up. Modern cosmologists are trying to improve the model using a period of very rapid expansion called inflation.

Evolution of the Universe

This page assumes the Standard Big Bang Model — so we can ignore newfangled theories like inflation (for now...)

Gravity Warps Space and Time

1) According to **general relativity** gravity works by changing the **shape** of space and time.

2) To reduce the brain-ache a bit, you can imagine the Universe as a **2-dimensional surface** that's warped in 3 dimensions. This is a handy way of getting an idea of what's going on, but **be careful**. **Space-time** actually has **4 dimensions** (x, y, z and time).

3) On a big scale, there are **three ways** that gravity can warp the Universe: the Universe can be **flat**, **open**, or **closed**.

4) This **curvature** of space-time determines the eventual **fate** of the Universe.

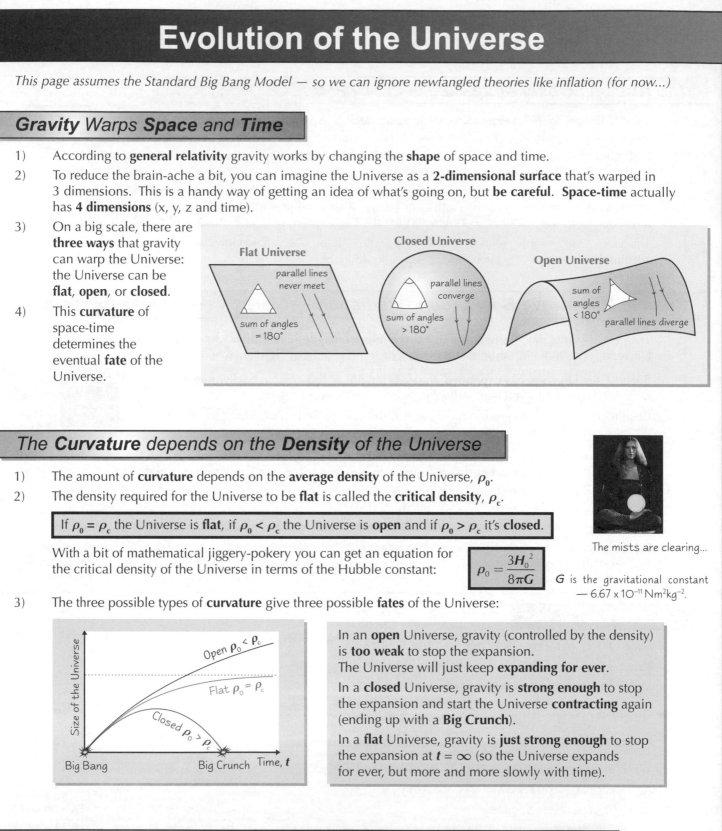

Flat Universe
parallel lines never meet
sum of angles = 180°

Closed Universe
parallel lines converge
sum of angles > 180°

Open Universe
sum of angles < 180°
parallel lines diverge

The Curvature depends on the Density of the Universe

1) The amount of **curvature** depends on the **average density** of the Universe, ρ_0.

2) The density required for the Universe to be **flat** is called the **critical density**, ρ_c.

> If $\rho_0 = \rho_c$ the Universe is **flat**, if $\rho_0 < \rho_c$ the Universe is **open** and if $\rho_0 > \rho_c$ it's **closed**.

With a bit of mathematical jiggery-pokery you can get an equation for the critical density of the Universe in terms of the Hubble constant:

$$\rho_0 = \frac{3H_0^2}{8\pi G}$$

The mists are clearing...

G is the gravitational constant — 6.67×10^{-11} Nm²kg⁻².

3) The three possible types of **curvature** give three possible **fates** of the Universe:

Size of the Universe (y-axis)
Open $\rho_0 < \rho_c$
Flat $\rho_0 = \rho_c$
Closed $\rho_0 > \rho_c$
Big Bang — Big Crunch — Time, t

In an **open** Universe, gravity (controlled by the density) is **too weak** to stop the expansion. The Universe will just keep **expanding for ever**.

In a **closed** Universe, gravity is **strong enough** to stop the expansion and start the Universe **contracting** again (ending up with a **Big Crunch**).

In a **flat** Universe, gravity is **just strong enough** to stop the expansion at $t = \infty$ (so the Universe expands for ever, but more and more slowly with time).

We Can't Calculate the Age of the Universe until we know its Density

1) A reasonable **estimate** of the **age** of the Universe is found from $t \approx 1/H_0$ (see page 156). But this formula assumes that the Universe had been expanding at the **same rate** for its whole lifetime.

2) In fact if you look at the **graph** of size against time, the **expansion rate** of the Universe is **slowing down**, even for the **open Universe**. So **in the past** the Universe was expanding **faster** than it is now.

3) That means we've **overestimated** the time it's taken for the Universe to get to the size it is now.

4) The **more dense** the Universe is, the **younger** it must be.

5) If you include all the "**dark matter**" and "**dark energy**" that's been detected **indirectly**, current **estimates** of the actual density of the Universe aren't very far off the **critical density**.

Evolution of the Universe

The *Story So Far...*

Before 10^{-4} seconds after the Big Bang, this is mainly guesswork. There are plenty of theories around, but not much experimental evidence to back them up. The general consensus at the moment goes something like this:

1) **Big Bang to 10^{-43} seconds.** Well, it's anybody's guess, really. At this sort of size and energy, even general relativity stops working properly. This is the "infinitely hot, infinitely small, infinitely dense" bit.

2) **10^{-43} seconds to 10^{-4} seconds.** At the start of this period, there's no distinction between different types of force — there's just one grand unified force. Then the Universe expands and cools, and the unified force splits into gravity, strong nuclear, weak nuclear and electromagnetic forces. Many cosmologists believe the Universe went through a rapid period of expansion called inflation at about 10^{-34} s.

   The Universe is a sea of quarks, antiquarks, leptons and photons. The quarks aren't bound up in particles like protons and neutrons, because there's too much energy around.

   At some point, matter-antimatter symmetry gets broken, so slightly more matter is made than antimatter. Nobody knows exactly how or when this happened, but most cosmologists like to put it as early as possible in the history of the Universe (before inflation, even).

Now we're onto more solid ground

3) **10^{-4} seconds.** This corresponds to a temperature of about 10^{12} K. The Universe is cool enough for quarks to join up to form particles like protons and neutrons. They can never exist separately again. Matter and antimatter annihilate each other, leaving a small excess of matter and huge numbers of photons (resulting in the cosmic background radiation that we observe today).

4) **About 100 seconds.** Temperature has cooled to 10^9 K. The Universe is similar to the interior of a star. Protons are cool enough to fuse to form helium nuclei.

5) **About 300 000 years.** Temperature has cooled to about 3000 K. The Universe is cool enough for electrons (that were produced in the first millisecond) to combine with helium and hydrogen nuclei to form atoms. The Universe becomes transparent since there are no free charges for the photons to interact with. This process is called recombination.

6) **About 14 billion years (now).** Temperature has cooled to about 3 K. Slight density fluctuations in the Universe mean that, over time, clumps of matter have been condensed by gravity into galactic clusters, galaxies and individual stars.

Practice Questions

Q1 What are the three possible fates of the Universe?

Q2 Why does the calculated age of the Universe depend on its density?

Q3 What is recombination?

Exam Questions

Q1 (a) Some cosmologists believe the Universe to be flat. What evidence is there to suggest that this is the case? [2 marks]

    (b) Explain what is meant by a flat universe in terms of its geometry, and its evolution. [3 marks]

Q2 The upper limit on Hubble's constant is 100 kms^{-1}Mpc^{-1} (1 Mpc = 3.09×10^{22} m).

    (a) Work out the average density of the Universe if the Universe is flat and H_0 is 100 kms^{-1}Mpc^{-1}. [4 marks]

    (b) Given that the mass of a hydrogen atom is 1.7×10^{-27} kg, calculate the average number of hydrogen atoms in every m^3 of space if the entire mass of the Universe is hydrogen. [2 marks]

Q3 Starting from the production of matter and antimatter, describe the evolution of the Universe (including its structure) up to the present day.
In your answer, you should use a clear, logical progression of ideas. [10 marks]

It's the end of the world as we know it...

Recently, astronomers have found evidence that the expansion rate is actually <u>accelerating</u> — *because of something called dark energy. That means the simple shapes of the graphs on the last page might be a long way from the true picture...*

Exponentials and Natural Logs

Mwah ha ha ha... you've hacked your way through the rest of the book and think you've finally got to the end of A2 Physics, but no, there's this tasty titbit of exam fun to go. You can get asked to look at and work out values from log graphs all over the shop, from astrophysics to electric field strength. And it's easy when you know how...

Many Relationships in Physics are Exponential

A fair few of the relationships you need to know about in A2 Physics are **exponential** — where the **rate of change** of a quantity is **proportional** to the **amount** of the quantity left. Here are just a few you should have met before (if they don't ring a bell, go have a quick read about them)...

Charge on a capacitor — the decay of charge on a capacitor is proportional to the amount of charge left on the capacitor:

$$Q = Q_o\, e^{(-t/RC)}$$ (see p. 121)

Radioactive decay — the rate of decay is proportional to the **number of nuclei left** to decay in a sample:

$$N = N_o\, e^{(-\lambda t)}$$ (see p. 131)

The **activity** of a radioactive sample behaves in the same way:

$$A = A_o\, e^{(-\lambda t)}$$ (see p. 131)

You can Plot Exponential Relations Using the Natural Log, ln

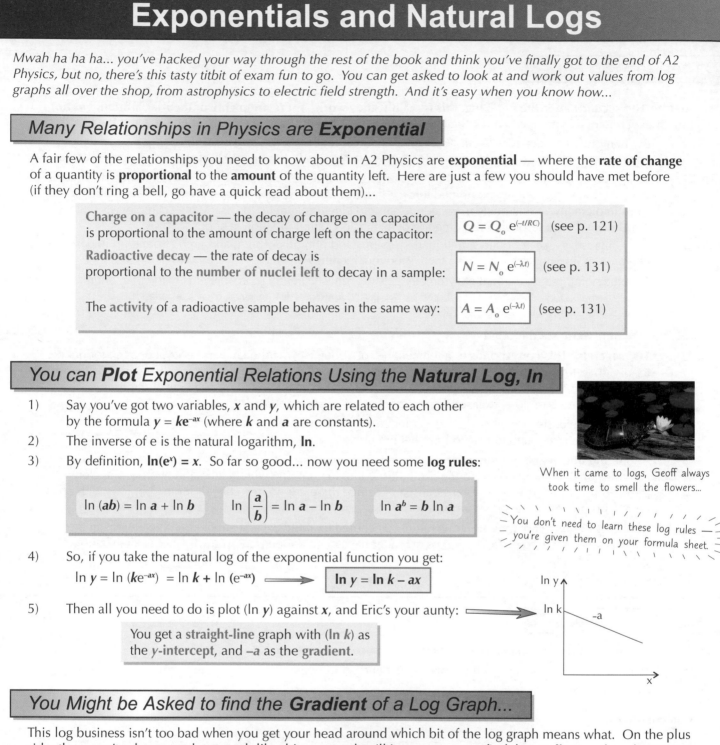

When it came to logs, Geoff always took time to smell the flowers...

1) Say you've got two variables, **x** and **y**, which are related to each other by the formula $y = ke^{-ax}$ (where **k** and **a** are constants).

2) The inverse of e is the natural logarithm, **ln**.

3) By definition, $\ln(e^x) = x$. So far so good... now you need some **log rules**:

$$\ln(ab) = \ln a + \ln b \qquad \ln\left(\frac{a}{b}\right) = \ln a - \ln b \qquad \ln a^b = b\ln a$$

You don't need to learn these log rules — you're given them on your formula sheet.

4) So, if you take the natural log of the exponential function you get:

$$\ln y = \ln(ke^{-ax}) = \ln k + \ln(e^{-ax}) \implies \boxed{\ln y = \ln k - ax}$$

5) Then all you need to do is plot (ln **y**) against **x**, and Eric's your aunty:

> You get a **straight-line** graph with (ln **k**) as the **y-intercept**, and **−a** as the **gradient**.

You Might be Asked to find the Gradient of a Log Graph...

This log business isn't too bad when you get your head around which bit of the log graph means what. On the plus side, they won't ask you to plot a graph like this *(yipee)* — they'll just want you to find the **gradient** or the **y-intercept**.

Example — finding the radioactive half-life of material X

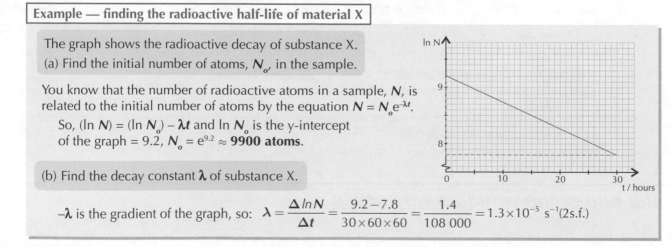

The graph shows the radioactive decay of substance X.

(a) Find the initial number of atoms, N_o, in the sample.

You know that the number of radioactive atoms in a sample, **N**, is related to the initial number of atoms by the equation $N = N_o e^{-\lambda t}$.
So, $(\ln N) = (\ln N_o) - \lambda t$ and $\ln N_o$ is the y-intercept of the graph = 9.2, $N_o = e^{9.2} \approx$ **9900 atoms**.

(b) Find the decay constant λ of substance X.

$-\lambda$ is the gradient of the graph, so: $\lambda = \dfrac{\Delta \ln N}{\Delta t} = \dfrac{9.2 - 7.8}{30 \times 60 \times 60} = \dfrac{1.4}{108\,000} = 1.3 \times 10^{-5}\ \text{s}^{-1}$ (2 s.f.)

Log Graphs and Long Answer Questions

You can Plot **Any Power Law** as a **Log-Log Graph**

You can use logs to plot a straight-line graph of **any power law** — it doesn't have to be an exponential.
Take the relationship between the energy stored in a spring, **E**, and the spring's extension, **x**:

$$E = kx^n$$

Take the log (base 10) of both sides to get:

$$\log E = \log k + n \log x$$

So **log k** will be the **y-intercept** and **n** the gradient of the graph.

Example

The graph shows how the intensity of radiation from the Sun, **I**, varies with its distance, **d**.
I is related to **d** by the power law **I = kd^n**. Find **n**.

$\log I = \log (kd^n) = \log k + \log d^n$
$= \log k + n \log d$.

so **n** is the **gradient** of the graph.
Reading from the graph:

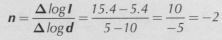

$$n = \frac{\Delta \log I}{\Delta \log d} = \frac{15.4 - 5.4}{5 - 10} = \frac{10}{-5} = -2$$

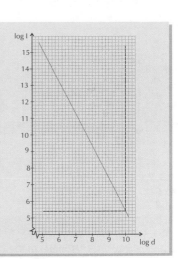

And that's the End of Logs... Now **Explain Yourself**...

In A2, they often give a couple of marks for 'the quality of written communication' when you're writing a slightly long answer (and not just pumping numbers into an equation).
You can pick up a couple of easy marks just by making sure that you do the things in the fetching blue box.

1) **Explain** your ideas or argument **clearly** as this is usually what you'll get a mark for. And make sure you **answer the question** being asked — it's dead easy to go off on a tangent. Like my mate Phil always says... have I ever told you about Phil? Well he...

2) Write in **whole sentences**.

3) Use **correct spelling**, **grammar** and **punctuation**.

4) Also check how many marks the question is worth. If it's only a two-marker, they don't want you to spend half an hour writing an essay about it.

Example

A large group of people walk across a footbridge. When the frequency of the group's footsteps is 1 Hz, the bridge noticeably oscillates and 'wobbles'.
Fully describe the phenomenon causing the bridge to wobble.
Suggest what engineers could do to solve this problem.
In your answer, you should make clear how your solution links with the theory. [6 marks]

Good Answer

The pedestrians provide a driving force on the bridge causing it to oscillate. At around 1 Hz, the driving frequency from the pedestrians is roughly equal to the natural frequency of the bridge, causing it to resonate. The amplitude of the bridge's oscillations when resonating at 1 Hz will be greater than at any other driving frequency. The oscillations at this frequency are large enough to be noticed by pedestrians.

Engineers could fix this problem by critically damping the bridge to stop any oscillations as quickly as possible.

They could also adjust the natural frequency of the bridge so that it was not so close to a known walking frequency of large groups of people.

Bad Answer

resonance
driving frequency of group = nat. freq.
damping

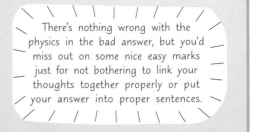

There's nothing wrong with the physics in the bad answer, but you'd miss out on some nice easy marks just for not bothering to link your thoughts together properly or put your answer into proper sentences.

Lumberjacks are great musicians — they have a natural logarithm...

Well, that's it folks. Crack open the chocolate bar of victory and know you've earnt it. Only the tiny detail of the actual exam to go... ahem. Make sure you know which bit means what on a log graph and you'll pick up some nice easy marks. Other than that, stay calm, be as clear as you can and good luck — I've got my fingers, toes and eyes crossed for you.

AS Answers

Unit 1: Section 1 — Motion

Page 5 — Scalars and Vectors

1) Start by drawing a diagram:

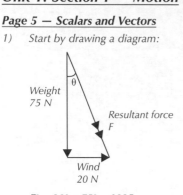

$F^2 = 20^2 + 75^2 = 6025$
So $F = 77.6$ N
$\tan\theta = 20/75 = 0.267$
So $\theta = 14.9°$
The resultant force on the rock is 77.6 N [1 mark]
at an angle of 14.9° [1 mark] to the vertical.

Make sure you know which angle you're finding — and label it on your diagram.

2) Again, start by drawing a diagram:

horizontal component $v_H = 20\cos15° = 19.3$ ms^{-1} [1 mark]
vertical component $v_V = 20\sin15° = 5.2$ ms^{-1} downwards [1 mark]
Always draw a diagram.

Page 7 — Motion with Constant Acceleration

1)a) $a = -9.81$ ms^{-2}, $t = 5$ s, $u = 0$ ms^{-1}, $v = ?$
use : $\quad v = u + at$
$\qquad v = 0 + 5 \times -9.81$ [1 mark for either step of working]
$\qquad v = -49.05$ ms^{-1} [1 mark]

NB: It's negative because she's falling downwards and we took upwards as the positive direction.

b) Use: $s = \left(\dfrac{u+v}{2}\right)t$ or $\quad s = ut + \frac{1}{2}at^2$ [1 mark for either]

$s = \dfrac{-49.05}{2} \times 5$ $\qquad s = 0 + \frac{1}{2} \times -9.81 \times 5^2$

$s = -122.625$ m $\qquad\qquad s = -122.625$ m
So she fell 122.625 m [1 mark for answer]

2)a) $v = 0$ ms^{-1}, $t = 3.2$ s, $s = 40$ m, $u = ?$

use: $s = \left(\dfrac{u+v}{2}\right)t$ [1 mark]

$40 = 3.2u \div 2$

$u = \dfrac{80}{3.2} = 25$ ms^{-1} [1 mark]

b) use: $v^2 = u^2 + 2as$ [1 mark]
$\quad 0 = 25^2 + 80a$
$-80a = 625$
$\quad a = -7.8$ ms^{-2} [1 mark]

3)a) Take upstream as negative: $v = 5$ ms^{-1}, $a = 6$ ms^{-2}, $s = 1.2$ m, $u = ?$
use: $v^2 = u^2 + 2as$ [1 mark]
$5^2 = u^2 + 2 \times 6 \times 1.2$
$u^2 = 25 - 14.4 = 10.6$
$u = -3.26$ ms^{-1} [1 mark]

b) From furthest point: $u = 0$ ms^{-1}, $a = 6$ ms^{-2}, $v = 5$ ms^{-1}, $s = ?$
use: $v^2 = u^2 + 2as$ [1 mark]
$5^2 = 0 + 2 \times 6 \times s$
$s = 25 \div 12 = 2.08$ m [1 mark]

Page 9 — Free Fall

1)a) The computer needs:
The time for the first strip of card to pass through the beam [1 mark]
The time for the second strip of card to pass through the beam [1 mark]
The time between these events [1 mark]

b) Average speed of first strip while it breaks the light beam = width of strip ÷ time to pass through beam [1 mark]
Average speed of second strip while it breaks the light beam = width of strip ÷ time to pass through beam [1 mark]
Acceleration = (second speed – first speed) ÷ time between light beam being broken [1 mark]

c) E.g. the device will accelerate while the beam is broken by the strips. [1 mark]

2)a) You know $s = 5$ m, $a = -g$, $v = 0$
You need to find u, so use $v^2 = u^2 + 2as$
$0 = u^2 - 2 \times 9.81 \times 5$ [1 mark for either line of working]
$u^2 = 98.1$, so $u = 9.9$ ms^{-1} [1 mark]

b) You know $a = -g$, $v = 0$ at highest pt, $u = 9.9$ ms^{-1} from a)
You need to find t, so use $v = u + at$
$0 = 9.9 - 9.81t$ [1 mark for either line of working]
$t = 9.9/9.81 = 1.0$ s [1 mark]

c) Her velocity as she lands back on the trampoline will be -9.9 ms^{-1} (same magnitude, opposite direction)
[2 marks — 1 for correct number, 1 for correct sign]

Page 11 — Free Fall and Projectile Motion

1)a) You only need to worry about the vertical motion of the stone.
$u = 0$ ms^{-1}, $s = -560$ m, $a = -g = -9.81$ ms^{-2}, $t = ?$
You need to find t, so use: $s = ut + \frac{1}{2}at^2$ [1 mark]
$-560 = 0 + \frac{1}{2} \times -9.81 \times t^2$

$t = \sqrt{\dfrac{2 \times (-560)}{-9.81}} = 10.7$ s (1 d.p.) $= 11$ s (to the nearest second)
[1 mark]

b) You know that in the horizontal direction:
$v = 20$ m/s, $t = 10.7$ s, $a = 0$, $s = ?$

So use velocity $= \dfrac{distance}{time}$, $v = \dfrac{s}{t}$ [1 mark]
$s = v \times t = 20 \times 10.7 = 214$ m (to the nearest metre) [1 mark]

AS Answers

2) You know that for the arrow's vertical motion (taking upwards as the positive direction):
$a = -9.81 \text{ ms}^{-2}$, $u = 30 \text{ ms}^{-1}$ and the arrow will be at its highest point just before it starts falling back towards the ground, so $v = 0 \text{ m/s}$.
s = the distance travelled from the arrow's firing point
So use $v^2 = u^2 + 2as$ [1 mark]
$0 = 30^2 + 2 \times -9.81 \times s$

$900 = 2 \times 9.81s$

$s = \dfrac{900}{2 \times 9.81} = 45.9 \text{ m}$ [1 mark]

So the maximum distance reached from the ground
$= 45.9 + 1 = 47 \text{ m}$ (to the nearest metre). [1 mark]

Page 13 — Displacement-Time Graphs

1) Split graph into four sections:

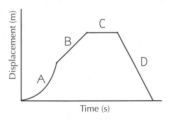

A: acceleration [1 mark]
B: constant velocity [1 mark]
C: stationary [1 mark]
D: constant velocity in opposite direction to A and B [1 mark]

2) a)

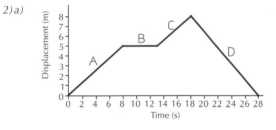

[4 marks — 1 mark for each section correctly drawn]

b) At A: $v = \dfrac{\text{displacement}}{\text{time}} = \dfrac{5}{8} = 0.625 \text{ ms}^{-1}$

At B: $v = 0$

At C: $v = \dfrac{\text{displacement}}{\text{time}} = \dfrac{3}{5} = 0.6 \text{ ms}^{-1}$

At D: $v = \dfrac{\text{displacement}}{\text{time}} = \dfrac{-8}{10} = -0.8 \text{ ms}^{-1}$

[2 marks for all correct or just 1 mark for 2 or 3 correct]

Page 15 — Velocity-Time Graphs

1) a)

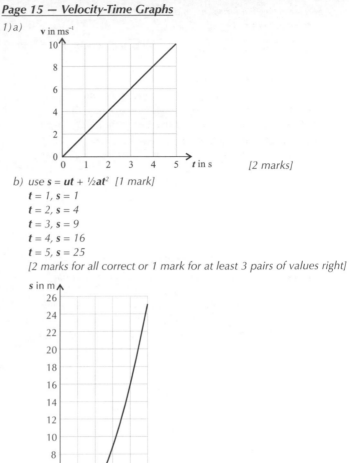

[2 marks]

b) use $s = ut + \frac{1}{2}at^2$ [1 mark]
$t = 1$, $s = 1$
$t = 2$, $s = 4$
$t = 3$, $s = 9$
$t = 4$, $s = 16$
$t = 5$, $s = 25$
[2 marks for all correct or 1 mark for at least 3 pairs of values right]

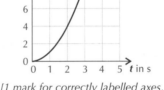
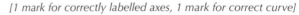

[1 mark for correctly labelled axes, 1 mark for correct curve]

c) E.g. another way to calculate displacement is to find the area under the velocity-time graph. [1 mark]
E.g. total displacement = $\frac{1}{2} \times 5 \times 10 = 25 \text{ m}$ [1 mark]

Unit 1: Section 2 — Forces in Action

Page 17 — Newton's Laws of Motion

1) a)   b)

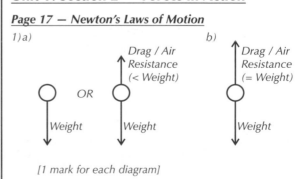

[1 mark for each diagram]

AS Answers

2) a) *Force perpendicular to river flow = 500 – 100 = 400 N [1 mark]*
   Force parallel to river flow = 300 N
   Magnitude of resultant force = $\sqrt{400^2 + 300^2}$ = 500 N [1 mark]

 b) *$a = F/m$ (from $F = ma$) [1 mark]*
   = 500/250 = 2 ms^{-2} [1 mark]

3) a) *The resultant force acting on it [1 mark] and its mass. [1 mark]*

 b) *E.g. Michael is able to exert a greater force than Tom.*
   Michael is lighter than Tom. [1 mark each for 2 sensible points]

 c) *The only force acting on each of them is their weight = mg*
   [1 mark]. Since $F = ma$, this gives $ma = mg$, or $a = g$ [1 mark].
   Their acceleration doesn't depend on their mass — it's the same
   for both of them — so they reach the water at the same time.
   [1 mark]

Page 19 — Drag and Terminal Velocity

1) a) *The velocity increases at a steady rate, which means the*
   acceleration is constant. [1 mark]
   Constant acceleration means there must be no atmospheric
   resistance (atmospheric resistance would increase with velocity,
   leading to a decrease in acceleration). So there must be no
   atmosphere. [1 mark]

 b)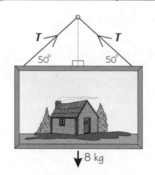
   [1 mark for a smooth curve that levels out, 1 mark for correct
   position relative to existing line]

   Your graph must be a smooth curve which levels out. It must NOT go
   down at the end.

 c) *(The graph becomes less steep)*
   because the acceleration is decreasing [1 mark]
   because air resistance increases with speed [1 mark]
   (The graph levels out)
   because air resistance has become equal to weight [1 mark]

   If the question says 'explain', you won't get marks for just describing
   what the graph shows — you have to say <u>why</u> it is that shape.

Page 21 — Mass, Weight and Centre of Gravity

1) a) *Density is a measure of 'compactness' of a material — its mass per*
   unit volume. [1 mark]

 b) *$\rho = \dfrac{m}{V}$ [1 mark]*
   V of cylinder = $\pi r^2 h = \pi \times 4^2 \times 6 = 301.6$ cm^3 [1 mark]
   $\rho = 820 \div 301.6 = 2.72$ g cm^{-3} [1 mark]

 c) *$V = 5 \times 5 \times 5 = 125$ cm^3*
   $m = \rho \times V = 2.7 \times 125 = 340$ g [1 mark]

2) *Experiment:*
   Hang the object freely from a point. Hang a plumb bob from the
   same point, and use it to draw a vertical line down the object.
   [1 mark]
   Repeat for a different point and find the point of intersection.
   [1 mark]
   The centre of gravity is halfway through the thickness of the object
   (by symmetry) at the point of intersection. [1 mark]
   Identifying and reducing error, e.g.:
   Source: the object and/or plumb line might move slightly while
   you're drawing the vertical line [1 mark]
   Reduced by: hang the object from a third point to confirm the
   position of the point of intersection [1 mark]

Page 23 — Forces and Equilibrium

1)
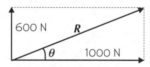

   Weight = vertical component of tension × 2
   $8 \times 9.81 = 2T \sin 50°$ [1 mark]
   $78.48 = 0.766 \times 2T$
   $102.45 = 2T$
   $T = 51.2$ N [1 mark]

2)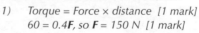

   By Pythagoras:
   $R = \sqrt{1000^2 + 600^2} = 1166$ N [1 mark]
   $\tan \theta = \dfrac{600}{1000}$, so $\theta = \tan^{-1} 0.6 = 31.0°$ [1 mark]

Page 25 — Moments and Torques

1) *Torque = Force × distance [1 mark]*
   $60 = 0.4F$, so $F = 150$ N [1 mark]

2)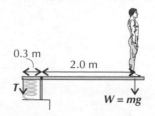

   clockwise moment = anticlockwise moment
   $W \times 2.0 = T \times 0.3$ [1 mark for either line of working]
   $60 \times 9.81 \times 2.0 = T \times 0.3$
   $T = 3924$ N [1 mark]

   The tension in the spring is equal and opposite to the force exerted by
   the diver on the spring.

AS Answers

Page 27 — Car Safety

1)a) reaction time is 0.5 s, speed is 20 ms^{-1}
$s = vt$ [1 mark]
$= 20 \times 0.5 = 10$ m [1 mark]

b) Use $F = ma$ to get a: $a = -10\,000/850 = -11.76$ ms^{-2} [1 mark]

Use $v^2 = u^2 + 2as$, and rearrange to get $s = \dfrac{v^2 - u^2}{2a}$

Put in the values: $s = (0 - 400) \div (2 \times -11.76)$ [1 mark]
$= 17$ m [1 mark]

Remember that a force against the direction of motion is negative.

c) Total stopping distance = 10 + 17 = 27 m
She stops 3 m before the cow. [1 mark]

2)a) Car: use $v = u + at$ to get acceleration:
$a = (0 - 20)/0.1 = -200$ ms^{-2} [1 mark]
Use $F = ma$:
$F = 900 \times -200 = -180\,000$ N [1 mark]
Same for dummy:
$a = 0 - 18/0.1 = -180$ ms^{-2} [1 mark]
$F = 50 \times -180 = -9000$ N [1 mark]

b) Crumple zones will increase the collision time for the car and dummy;
this reduces forces on the car and dummy;
the airbag will keep the dummy in its seat;
and increase the collision time further for the dummy;
reducing the force on it.
[3 marks for any three sensible points]

Unit 1: Section 3 — Work and Energy

Page 29 — Work and Power

1)a)

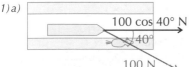

Force in direction of travel = 100 cos40° = 76.6 N [1 mark]
$W = Fs = 76.6 \times 1500 = 114\,900$ J [1 mark]

b) Use $P = Fv$ [1 mark]
= 100 cos40° × 0.8 = 61.3 W [1 mark]

2)a) Use $W = Fs$ [1 mark]
$= 20 \times 9.81 \times 3 = 588.6$ J [1 mark]

Remember that 20 kg is not the force — it's the mass. So you need to multiply it by 9.81 Nkg^{-1} to get the weight.

b) Use $P = Fv$ [1 mark]
$= 20 \times 9.81 \times 0.25 = 49.05$ W [1 mark]

Page 31 — Conservation of Energy

1)a) Use $E_k = \frac{1}{2}mv^2$ and $E_p = mgh$ [1 mark]
$\frac{1}{2}mv^2 = mgh$
$\frac{1}{2}v^2 = gh$
$v^2 = 2gh = 2 \times 9.81 \times 2 = 39.24$ [1 mark]
$v = 6.26$ ms^{-1} [1 mark]

'No friction' allows you to say that the changes in kinetic and potential energy will be the same.

b) 2 m — no friction means the kinetic energy will all change back into potential energy, so he will rise back up to the same height as he started. [1 mark]

c) Put in some more energy by actively 'skating'. [1 mark]

2)a) If there's no air resistance, $E_k = E_p = mgh$ [1 mark]
$E_k = 0.02 \times 9.81 \times 8 = 1.57$ J [1 mark]

b) If the ball rebounds to 6.5 m, it has gravitational potential energy:
$E_p = mgh = 0.02 \times 9.81 \times 6.5 = 1.28$ J [1 mark]
So 1.57 − 1.28 = 0.29 J is converted to other forms [1 mark]

Page 33 — Efficiency and Sankey Diagrams

1)a) Wasted heat energy = 125 − 30 − 70 = 25 KJ [1 mark]

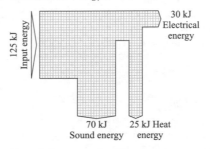

[1 mark for correctly drawn input arrow, 1 mark for correctly drawn output arrows, 1 mark for labels]

b) Efficiency = $\dfrac{\text{useful output energy}}{\text{total input energy}} \times 100\%$

Efficiency of first design = $\dfrac{15}{60} \times 100\% = 25\%$ [1 mark]

Efficiency of second design = $\dfrac{30}{125} \times 100\% = 24\%$ [1 mark]

The first design is 1% more efficient than the second. [1 mark]

Page 35 — Hooke's Law

1)a) Force is proportional to extension.
The force is 1.5 times as great, so the extension will also be 1.5 times the original value.
Extension = 1.5 × 4.0 mm = 6.0 mm [1 mark]

b) $F = ke$ and so $k = F/e$ [1 mark]
$k = 10 \div 4.0 \times 10^{-3} = 2500$ Nm^{-1} or 2.5 Nmm^{-1} [1 mark]
There is one mark for rearranging the equation and another for getting the right numerical answer.

c) One mark for any sensible point e.g.
The string now stretches much further for small increases in force.
When the string is loosened it is longer than at the start. [1 mark]

2) The rubber band does not obey Hooke's law [1 mark]
because when the force is doubled from 2.5 N to 5 N, the extension increases by a factor of 2.3. [1 mark]

AS Answers

Page 37 — Stress and Strain

1)a) Area = $\pi d^2/4$ or πr^2.
   So area = $\pi \times (1 \times 10^{-3})^2/4 = 7.85 \times 10^{-7}$ m² [1 mark]

   b) Stress = force/area = $300/(7.85 \times 10^{-7}) = 3.82 \times 10^8$ Nm⁻² [1 mark]

   c) Strain = extension/length = $4 \times 10^{-3}/2.00 = 2 \times 10^{-3}$ [1 mark]

2)a) $F = ke$ and so rearranging $k = F/e$ [1 mark]
   So $k = 50/(3.0 \times 10^{-3}) = 1.67 \times 10^4$ Nm⁻¹ [1 mark]

   b) Elastic strain energy = ½Fe
   Giving the elastic strain energy as
   ½ $\times 50 \times 3 \times 10^{-3} = 7.5 \times 10^{-2}$ J [1 mark]

3) Elastic strain energy,
   E = ½ke^2 = ½ $\times 40.8 \times 0.05^2 = 0.051$ J [1 mark]
   To find maximum speed, assume all this energy is converted to kinetic energy in the ball. $E_{kinetic} = E$ [1 mark]
   E = ½mv^2, so rearranging, $v^2 = 2E/m$ [1 mark]
   $v^2 = (2 \times 0.051)/0.012 = 8.5$, so $v = 2.92$ ms⁻¹ [1 mark]

Page 39 — The Young Modulus

1)a) Cross-sectional area = $\pi d^2/4$ or πr^2.
   So the cross-sectional area = $\pi \times (0.6 \times 10^{-3})^2/4 = 2.83 \times 10^{-7}$ m² [1 mark]

   b) Stress = force/area = $80/(2.83 \times 10^{-7}) = 2.83 \times 10^8$ Nm⁻² [1 mark]

   c) Strain = extension/length = $3.6 \times 10^{-3}/2.5 = 1.44 \times 10^{-3}$ [1 mark]

   d) The Young modulus for steel = stress/strain
   = $2.83 \times 10^8/(1.44 \times 10^{-3}) = 2.0 \times 10^{11}$ Nm⁻² [1 mark]

2)a) The Young modulus, E = stress/strain and so strain = stress/E [1 mark]
   Strain on copper = $2.6 \times 10^8/1.3 \times 10^{11} = 2 \times 10^{-3}$ [1 mark]
   There's 1 mark for rearranging the equation and another for using it.

   b) Stress = force/area and so area = force/stress
   Area of the wire = $100/(2.6 \times 10^8) = 3.85 \times 10^{-7}$ m² [1 mark]

   c) Strain energy per unit volume = ½ $\times$ stress $\times$ strain
   = ½ $\times 2.6 \times 10^8 \times 2 \times 10^{-3} = 2.6 \times 10^5$ Jm⁻³ [1 mark]
   Give the mark if answer is consistent with the value calculated for strain in part a).

Page 41 — Interpreting Stress-Strain Graphs

1)a) Liable to break suddenly without deforming plastically. [1 mark]

   b)
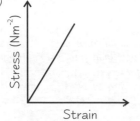
   [1 mark for correctly labelled axes, 1 mark for straight line through the origin]

2)a)
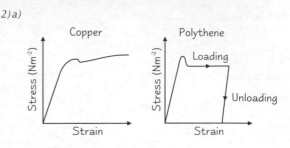
[1 mark for correctly labelled axes, 1 mark for correct shape of copper graph, 1 mark for correct shape of polythene graph (loading curve only is acceptable]

   b) The stress-strain graphs for both materials begin with a straight line through the origin, showing that both materials initially obey Hooke's law. [1 mark]
   Both materials undergo plastic deformation when a large enough stress is applied. [1 mark]

Unit 2: Section 1 — Electric Circuits, Resistance and DC Circuits

Page 43 — Charge, Current and Potential Difference

1) Time in seconds = $10 \times 60 = 600$ s.
   Use the formula $I = Q / t$ [1 mark]
   which gives you $I = 4500 / 600 = 7.5$ A [1 mark]
   Write down the formula first. Don't forget the unit in your answer.

2) Rearrange the formula $I = nAvq$ and you get $v = I / nAq$ [1 mark]
   which gives you
   $$v = \frac{13}{(1.0 \times 10^{29}) \times (5.0 \times 10^{-6}) \times (1.6 \times 10^{-19})}$$ [1 mark]
   $v = 1.63 \times 10^{-4}$ ms⁻¹ [1 mark]

3) Work done = $0.75 \times$ electrical energy input
   so the energy input will be $90 / 0.75 = 120$ J. [1 mark]
   Rearrange the formula $V = W / Q$ to give $Q = W / V$ [1 mark]
   so you get $Q = 120 / 12 = 10$ C. [1 mark]
   The electrical energy input to a motor has to be greater than the work it does because motors are less than 100% efficient.

Page 45 — Resistance and Resistivity

1) Area = $\pi(d/2)^2$ and $d = 1.0 \times 10^{-3}$ m
   so Area = $\pi \times (0.5 \times 10^{-3})^2 = 7.85 \times 10^{-7}$ m² [1 mark]
   $$R = \frac{\rho l}{A} = \frac{2.8 \times 10^{-8} \times 4}{7.85 \times 10^{-7}} = 0.14 \ \Omega$$
   [1 mark for equation or working, 1 mark for answer with unit.]

2)a) $R = V / I$ [1 mark]
   $$= \frac{2}{2.67 \times 10^{-3}} = 749 \ \Omega$$ [1 mark]

   b) Two further resistance calculations give 750 Ω for each answer [1 mark]
   There is no significant change in resistance for different potential differences [1 mark]
   Component is an ohmic conductor because its resistance is constant for different potential differences. [1 mark]

AS Answers

Page 47 — I/V Characteristics

1) a)

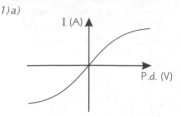

[1 mark]

b) Resistance increases as the temperature increases *[1 mark]*

c) Increase in temperature makes metal ions vibrate more *[1 mark]*
   Increased collisions with ions impedes electrons *[1 mark]*

Page 49 — Electrical Energy and Power

1) a) $I = P/V$ *[1 mark]* = $920/230 = 4$ A *[1 mark]*

b) $I = V/R$ *[1 mark]* = $230/190 = 1.21$ A *[1 mark]*

c) $P_{motor} = VI = 230 \times 1.21 = 278$ W *[1 mark]*
   Total power = motor power + heater power
   = $278 + 920 = 1198$ W, which is approx. 1.2 kW *[1 mark]*

2) a) Energy supplied = $VIt = 12 \times 48 \times 2 = 1152$ J *[1 mark]*

b) Energy lost = I^2Rt *[1 mark]* = $48^2 \times 0.01 \times 2 = 46$ J *[1 mark]*

Page 51 — Domestic Energy and Fuses

1) a) Energy = power × time
   i) Energy = $1800 \times (15 \times 60) = 1\ 620\ 000$ J (= 1.62 MJ) *[1 mark]*
   ii) Energy = $1.8 \times (15 \div 60) = 0.45$ kWh *[1 mark]*

b) Cost = number of units × price per unit = $0.45 \times 14.6 = 6.57$p
   [1 mark]

2) a) $P = VI$. Rearranging, $I = P/V = 1500 \div 230 = 6.52$ A *[1 mark]*
   A 13 A fuse should be used. *[1 mark]*

b) $E = Pt = 1.5 \times 2.25 = 3.375$ kWh *[1 mark]*
   Cost = $3.375 \times 9.8 = 33.075$p *[1 mark]*

c) $P = VI = 230 \times (6.5 \times 10^{-3}) = 1.495 \approx 1.5$ W *[1 mark]*

d) $E = Pt = 0.0015 \times 10 = 0.015$ kWh *[1 mark]*
   Cost = $0.015 \times 9.8 = 0.147$p *[1 mark]*

Page 53 — E.m.f. and Internal Resistance

1) a) Total resistance = $R + r = 4 + 0.8 = 1.8$ Ω *[1 mark]*
   I = e.m.f./total resistance = $24/4.8 = 5$ A *[1 mark]*

b) $V = \varepsilon - Ir = 24 - 5 \times 0.8 = 20$ V *[1 mark]*

2) a) $\varepsilon = I(R + r)$, so $r = \varepsilon/I - R$ *[1 mark]*
   $r = 500/(50 \times 10^{-3}) - 10 = 9990$ Ω *[1 mark]*

b) This is a very high internal resistance *[1 mark]*
   So only small currents can be drawn, reducing the risk to the user
   [1 mark]

Page 55 — Conservation of Energy & Charge in Circuits

1) a) Resistance of parallel resistors:
   $1/R_{parallel} = 1/6 + 1/3 = 1/2$
   $R_{parallel} = 2$ Ω *[1 mark]*
   Total resistance:
   $R_{total} = 4 + R_{parallel} = 4 + 2 = 6$ Ω *[1 mark]*

b) $V = IR$, so rearranging $I_3 = V / R_{total}$ *[1 mark]*
   $I_3 = 12 / 6 = 2$ A *[1 mark]*

c) $V = IR = 2 \times 4 = 8$ V *[1 mark]*

d) E.m.f. = sum of p.d.s in circuit, so $12 = 8 + V_{parallel}$
   $V_{parallel} = 12 - 8 = 4$ V *[1 mark]*

e) Current = p.d. / resistance
   $I_1 = 4 / 3 = 1.33$ A *[1 mark]*
   $I_2 = 4 / 6 = 0.67$ A *[1 mark]*

Page 57 — The Potential Divider

1) Parallel circuit, so p.d. across both sets of resistors is 12 V.
   i) $V_{AB} = \frac{1}{2} \times 12 = 6$ V *[1 mark]*
   ii) $V_{AC} = 2/3 \times 12 = 8$ V *[1 mark]*
   iii) $V_{BC} = V_{AC} - V_{AB} = 8 - 6 = 2$ V *[1 mark]*

2) a) $V_{AB} = 50/80 \times 12 = 7.5$ V *[1 mark]*
   (ignore the 10 Ω — no current flows that way)

b) Total resistance of the parallel circuit:
   $1/R_T = 1/50 + 1/(10 + 40) = 1/25$
   $R_T = 25$Ω *[1 mark]*
   p.d. over the whole parallel arrangement = $25/55 \times 12 = 5.45$ V
   [1 mark]
   p.d. across AB = $40/50 \times 5.45 = 4.36$ V *[1 mark]*
   current through 40 Ω resistor = $V/R = 4.36/40 = 0.11$ A *[1 mark]*

Unit 2: Section 2 — Waves

Page 59 — The Nature of Waves

1) a) Use $v = \lambda f$ and $f = 1 / T$
   So $v = \lambda / T$, giving $\lambda = vT$ *[1 mark]*
   $\lambda = 3$ ms^{-1} × 6 s = 18 m *[1 mark]*
   The vertical movement of the buoy is irrelevant to this part of the
   question

b) The trough to peak distance is twice the amplitude, so the
   amplitude is 0.6 m *[1 mark]*

Page 61 — Longitudinal and Transverse Waves

1) a) [This question could equally well be answered using diagrams.]
   For ordinary light, the EM field vibrates in all planes at right angles
   to the direction of travel. *[1 mark]*
   Iceland spar acts as a polariser. *[1 mark]*
   When light is shone through the first disc, it only allows through
   vibrations in one particular plane, so emerges less bright. *[1 mark]*
   As the two crystals are rotated relative to each other there comes a
   point when the allowed planes are at right angles to each other.
   [1 mark] So all the light is blocked. *[1 mark]*
   Try to remember to say that for light and other EM waves it's the
   electric and magnetic <u>fields</u> that vibrate.

AS Answers

b) Using Malus' law, $I = I_0 \cos^2 \sigma$ [1 mark]. I is half the size of I_0, so $I/I_0 = 0.5$ [1 mark]. Which means $\cos^2\sigma = 0.5$, so $\cos\sigma = 0.707...$ and $\sigma = 45°$ [1 mark]

2) E.g. Polarising filters are used in photography to remove unwanted reflections [1 mark]. Light is partially polarised when it reflects so putting a polarising filter over the lens at 90 degrees to the plane of polarisation will block most of the reflected light. [1 mark].

Page 63 — The Electromagnetic Spectrum

1) At the same speed. [1 mark]
Both are electromagnetic waves and hence travel at **c** in a vacuum. [1 mark]

2) a) Medical X-rays [1 mark] rely on the fact that X-rays penetrate the body well but are blocked by bone. [1 mark]
OR
Security scanners at airports [1 mark] rely on the fact that X-rays penetrate suitcases and clothes but are blocked by metal e.g. of a weapon. [1 mark]

b) The main difference between gamma rays and X-rays is that gamma rays arise from nuclear decay [1 mark] but X-rays are generated when metals are bombarded with electrons. [1 mark]

3) Any of: unshielded microwaves, excess heat, damage to eyes from too bright light, sunburn or skin cancer from UV, cancer or eye damage due to ionisation by X-rays or gamma rays.
[1 mark for the type of EM wave, 1 mark for the danger to health]

Page 65 — Superposition and Coherence

1) a) The frequencies and wavelengths of the two sources must be equal [1 mark] and the phase difference must be constant. [1 mark]

b) Interference will only be noticeable if the amplitudes of the two waves are approximately equal. [1 mark]

2) a) 180° (or 180° + 360n°). [1 mark]

b) The displacements and velocities of the two points are equal in size [1 mark] but in opposite directions. [1 mark]

Page 67 — Standing (Stationary) Waves

1) a)

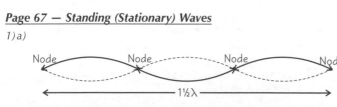

[1 mark for the correct shape, 1 mark for labelling the length]

b) For a string vibrating at three times the fundamental frequency, length = $3\lambda / 2$
$1.2\ m = 3\lambda / 2$
$\lambda = 0.8\ m$ [1 mark]

c) When the string forms a standing wave, its amplitude varies from a maximum at the antinodes to zero at the nodes. [1 mark] In a progressive wave all the points have the same amplitude. [1 mark]

Page 69 — Diffraction

1) When a wavefront meets an obstacle, the waves will diffract round the corners of the obstacle. When the obstacle is much bigger than the wavelength, little diffraction occurs. In this case, the mountain is much bigger than the wavelength of short-wave radio. So the "shadow" where you cannot pick up short wave is very long. [1 mark]

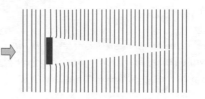

[1 mark]

When the obstacle is comparable in size to the wavelength, as it is for the long-wave radio waves, more diffraction occurs. The wavefront re-forms after a shorter distance, leaving a shorter "shadow". [1 mark]

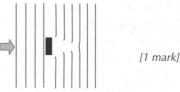

[1 mark]

Page 71 — Two-Source Interference

1) a)

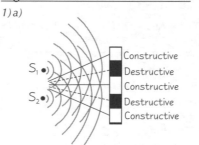

[1 mark for correct constructive interference patterns, 1 mark for correct destructive interference patterns]

b) Light waves from separate sources are not coherent, as light is emitted in random bursts of energy. To get coherent light the two sets of waves must emerge from one source. [1 mark] A laser is used because it emits coherent light that is all of one wavelength. [1 mark]

2) a) $\lambda = v/f = 330 / 1320 = 0.25\ m$. [1 mark]

b) Separation = $X = \lambda D / d$ [1 mark]
$= 0.25\ m \times 7\ m / 1.5\ m = 1.17\ m$. [1 mark]

AS Answers

Page 73 — Diffraction Gratings

1) a) Use $\sin\theta = n\lambda / d$

For the first order, $n = 1$

So, $\sin\theta = \lambda / d$ [1 mark]

No need to actually work out d. The number of lines per metre is 1 / d. So you can simply multiply the wavelength by that.

$\sin\theta = 600 \times 10^{-9} \times 4 \times 10^5 = 0.24$

$\theta = 13.9°$ [1 mark]

For the second order, $n = 2$ and $\sin\theta = 2\lambda / d$. [1 mark]

You already have a value for λ / d. Just double it to get $\sin\theta$ for the second order.

$\sin\theta = 0.48$

$\theta = 28.7°$ [1 mark]

b) No. Putting $n = 5$ into the equation gives a value of $\sin\theta$ of 1.2, which is impossible. [1 mark]

2) $\sin\theta = n\lambda / d$, so for the 1st order maximum, $\sin\theta = \lambda / d$ [1 mark]

$\sin 14.2° = \lambda \times 3.7 \times 10^5$

$\lambda = 663$ nm (or 6.63×10^{-7} m) [1 mark].

Unit 2: Section 3 — Quantum Phenomena

Page 75 — Light — Wave or Particle

1) a) At threshold voltage: $E_{kinetic}$ of an electron $= E_{photon}$ emitted [1 mark]

So $E_{photon} = e \times V = 1.6 \times 10^{-19} \times 1.7 = 2.72 \times 10^{-19}$ J [1 mark]

b) $E = \dfrac{hc}{\lambda}$ [1 mark], so $h = \dfrac{E\lambda}{c}$

$\lambda = 7.0 \times 10^{-7}$ m, $c = 3.0 \times 10^8$ ms^{-1},

So, $h = \dfrac{2.72 \times 10^{-19} \times 7.0 \times 10^{-7}}{3.0 \times 10^8} = 6.3 \ 10^{-34}$ Js [1 mark]

Page 77 — The Photoelectric Effect

1) $\phi = 2.9$ eV $= 2.9 \times (1.6 \times 10^{-19})$ J $= 4.64 \times 10^{-19}$ J [1 mark]

$f = \dfrac{\phi}{h} = \dfrac{4.64 \times 10^{-19}}{6.6 \times 10^{-34}} = 7.0 \times 10^{14}$ Hz (to 2 s.f.) [1 mark]

2) a) $E = hf$ [1 mark]

$= (6.6 \times 10^{-34}) \times (2.0 \times 10^{15}) = 1.32 \times 10^{-18}$ J [1 mark]

1.32×10^{-18} J $= \dfrac{1.32 \times 10^{-18}}{1.6 \times 10^{-19}}$ eV $= 8.25$ eV [1 mark]

b) $E_{photon} = E_{max\ kinetic} + \phi$ [1 mark]

$E_{max\ kinetic} = E_{photon} - \phi = 8.25 - 4.7 = 3.55$ eV (or 5.68×10^{-19} J) [1 mark]

3) An electron needs to gain a certain amount of energy (the work function energy) before it can leave the surface of the metal [1 mark]

If the energy carried by each photon is less than this work function energy, no electrons will be emitted [1 mark].

Page 79 — Energy Levels and Photon Emission

1) a) i) $E = V = 12.1$ eV [1 mark]

ii) $E = V \times 1.6 \times 10^{-19} = 12.1 \times 1.6 \times 10^{-19} = 1.9 \times 10^{-18}$ J [1 mark]

b) i) Excitation occurs when an electron moves from a lower energy to a higher energy level by absorbing energy. [1 mark]

ii) $-13.6 + 12.1 = -1.5$ eV. This corresponds to $n = 3$. [1 mark]

iii) $n = 3 \rightarrow n = 2$: $3.4 - 1.5 = 1.9$ eV [1 mark]

$n = 2 \rightarrow n = 1$: $13.6 - 3.4 = 10.2$ eV [1 mark]

$n = 3 \rightarrow n = 1$: $13.6 - 1.5 = 12.1$ eV [1 mark]

Page 81 — Wave-Particle Duality

1) a) Electromagnetic radiation can show characteristics of both a particle and a wave. [1 mark]

b) i) $E_{photon} = \dfrac{hc}{\lambda} = \dfrac{6.63 \times 10^{-34} \times 3.00 \times 10^8}{590 \times 10^{-9}}$ [1 mark]

$= 3.37 \times 10^{-19}$ J [1 mark]

$E\ (in\ eV) = \dfrac{E\ (in\ J)}{1.6 \times 10^{-19}} = \dfrac{3.37 \times 10^{-19}}{1.6 \times 10^{-19}} = 2.11$ eV [1 mark]

ii) $\lambda = \dfrac{h}{mv}$

$\therefore v = \dfrac{h}{m\lambda} = \dfrac{6.63 \times 10^{-34}}{9.1 \times 10^{-31} \times 590 \times 10^{-9}} = 1230$ ms^{-1}

[2 marks, otherwise 1 mark for some correct working]

2) a) $\lambda = \dfrac{h}{mv} = \dfrac{6.63 \times 10^{-34}}{9.1 \times 10^{-31} \times 3.5 \times 10^6} = 2.08 \times 10^{-10}$ m

[2 marks, otherwise 1 mark for some correct working]

b) Either $v = \dfrac{h}{m\lambda}$ with $m_{proton} = 1840 \times m_{electron}$

or momentum of protons = momentum of electrons

$1840 \times \cancel{m_e} \times v_p = \cancel{m_e} \times 3.5 \times 10^6$

$v_p = 1900$ ms^{-1}

[2 marks, otherwise 1 mark for some correct working]

c) The two have the same kinetic energy if the voltages are the same. The proton has a larger mass, so it will have a smaller speed. [1 mark] Kinetic energy is proportional to the square of the speed, while momentum is proportional to the speed, so they will have different momenta. [1 mark]

Wavelength depends on the momentum, so the wavelengths are different. [1 mark]

This is a really hard question. If you didn't get it right, make sure you understand the answer fully. Do the algebra if it helps.

3) a) $E_k = 6 \times 10^3$ eV [1 mark]

$= 6000 \times 1.6 \times 10^{-19} = 9.6 \times 10^{-16}$ J [1 mark]

b) $E_k = \dfrac{1}{2}mv^2$

$9.6 \times 10^{-16} = \dfrac{1}{2} \times 9.1 \times 10^{-31} \times v^2$

$v = \sqrt{\dfrac{2 \times 9.6 \times 10^{-16}}{9.1 \times 10^{-31}}} = 4.6 \times 10^7$ ms^{-1}

[2 marks, otherwise 1 mark for some correct working]

c) $\lambda = \dfrac{h}{mv} = \dfrac{6.63 \times 10^{-34}}{9.1 \times 10^{-31} \times 4.6 \times 10^7} = 1.58 \times 10^{-11}$ m

[2 marks, otherwise 1 mark for some correct working]

A2 Answers

Unit 4: Section 1 — Newton's Laws and Momentum

Page 87 — Momentum and Impulse

1) a) total momentum before collision = total momentum after [1 mark]
$(0.6 \times 5) + 0 = (0.6 \times -2.4) + 2v$
$3 + 1.44 = 2v$ [1 mark fo r working] $\Rightarrow v = 2.22 \ ms^{-1}$ [1 mark]

b) Kinetic energy before collision = ½ × 0.6 × 5² + ½ × 2 × 0² = 7.5 J
Kinetic energy after the collision = ½ × 0.6 × 2.4² + ½ × 2 × 2.22²
= 1.728 + 4.9284 = 6.6564 J [1 mark]
The kinetic energy of the two balls is greater before the collision
than after (i.e. it's not conserved) [1 mark], so the collision must
be inelastic [1 mark].

2) momentum before = momentum after [1 mark]
$(0.7 \times 0.3) + 0 = 1.1v$
$0.21 = 1.1v$ [1 mark for working] $\Rightarrow v = 0.19 \ ms^{-1}$ [1 mark]

Page 89 — Newton's Laws of Motion

1) a) When the parachutist first jumps out of the plane, the only vertical
force acting on her is due to gravity, so there is a resultant
downward force [1 mark]. Newton's 2nd law states that the
acceleration of a body is proportional to the resultant force, so she
will accelerate downwards [1 mark].

b) $F = ma = mg = 78 \times 9.81 = 765.18 \ N$ [1 mark]

c) Newton's 1st law states that a force is needed to change the
velocity of an object [1 mark] — the parachutist's velocity is not
changing,
so the resultant force acting on her must be zero [1 mark].

2) Force perpendicular to river flow = 500 – 100 = 400 N [1 mark]
Force parallel to river flow = 300 N
Resultant force = $\sqrt{400^2 + 300^2} = 500 \ N$ [1 mark]
$a = F/m$ [1 mark] = 500/250 = 2 ms⁻² [1 mark]

Unit 4: Section 2 — Circular Motion and Oscillations

Page 91 — Circular Motion

1) a) $\omega = \dfrac{\theta}{t}$ [1 mark] so $\omega = \dfrac{2\pi}{3.2 \times 10^7} = 2.0 \times 10^{-7} \ rad \, s^{-1}$ [1 mark]

b) $v = r\omega$ [1 mark] = 1.5 × 10¹¹ × 2.0 × 10⁻⁷ = 30 kms⁻¹ [1 mark]

c) $F = m\omega^2 r$ [1 mark] = 6.0 × 10²⁴ × (2.0 × 10⁻⁷)² × 1.5 × 10¹¹
= 3.6 × 10²² N [1 mark]

The answers to b) and c) use the rounded value of ω calculated in
part a) — if you didn't round, your answers will be slightly different.

d) The gravitational force between the Sun and the Earth [1 mark]

2) a) Gravity pulling down on the water at the top of the swing gives a
centripetal acceleration of 9.81 ms⁻² [1 mark]. If the circular
motion of the water needs a centripetal acceleration of less than
9.81 ms⁻², gravity will pull it in too tight a circle. The water will fall
out of the bucket.

Since $a = \omega^2 r$, $\omega^2 = \dfrac{a}{r} = \dfrac{9.81}{1}$, so $\omega = 3.1 \ rad \, s^{-1}$ [1 mark]

$\omega = 2\pi f$, so $f = \dfrac{\omega}{2\pi} = 0.5 \ rev \, s^{-1}$ [1 mark]

b) Centripetal force = $m\omega^2 r$ = 10 × 5² × 1 = 250 N [1 mark].
This force is provided by both the tension in the rope, T, and
gravity:
$T + (10 \times 9.81) = 250$. So $T = 250 - (10 \times 9.81) = 152 \ N$
[1 mark].

Page 93 — Gravitational Fields

1) $g = \dfrac{GM}{r^2} \Rightarrow M = \dfrac{gr^2}{G} = \dfrac{9.81 \times (6400 \times 1000)^2}{6.67 \times 10^{-11}}$ [1 mark]
$= 6.02 \times 10^{24} \ kg$ [1 mark]

2) a) $g = \dfrac{GM}{r^2} = \dfrac{6.67 \times 10^{-11} \times 7.35 \times 10^{22}}{(1740 \times 1000)^2} = 1.62 \ Nkg^{-1}$ [1 mark]

b) $F = -\dfrac{GMm}{r^2} = \dfrac{6.67 \times 10^{-11} \times 7.35 \times 10^{22} \times 25}{(1740 \times 1000 + 10)^2}$ [1 mark]
$= 40.48 \ N$ [1 mark]

Page 95 — Motion of Masses in Gravitational Fields

1) a) $T = \sqrt{\dfrac{4\pi^2 r^3}{GM}} = \sqrt{\dfrac{4\pi^2 \times [(6400 + 200) \times 1000]^3}{6.67 \times 10^{-11} \times 5.98 \times 10^{24}}}$ [1 mark]
$= 5334 \ seconds \ OR \ 1.48 \ hours$ [1 mark]

b) $v = \sqrt{\dfrac{GM}{r}} = \sqrt{\dfrac{6.67 \times 10^{-11} \times 5.98 \times 10^{24}}{(6400 + 200) \times 1000}}$
$= 7774 \ ms^{-1} \approx 7.77 \ kms^{-1}$ [1 mark]

2) Period = 24 hours = 24 × 60 × 60 = 86 400 s [1 mark]

$T = \sqrt{\dfrac{4\pi^2 r^3}{GM}} \Rightarrow r = \sqrt[3]{\dfrac{T^2 GM}{4\pi^2}} = \sqrt[3]{\dfrac{86400^2 \times 6.67 \times 10^{-11} \times 5.98 \times 10^{24}}{4\pi^2}}$

$r = 4.23 \times 10^7 \ m = 4.23 \times 10^4 \ km$ [1 mark]
Height above Earth = 4.23 × 10⁴ – 6.4 × 10³ = 35 900 km [1 mark]

3) Over 50 000 years, the Sun will have only lost a tiny fraction of its
mass (9.5 × 10²¹kg overall) [1 mark], which will not have caused
any significant change in the Earth's orbit [1 mark].

Page 97 — Simple Harmonic Motion

1) a) Simple harmonic motion is an oscillation in which an object always
accelerates towards a fixed point [1 mark] with an acceleration
directly proportional to its displacement from that point [1 mark].
[The SHM equation would get you the marks if you defined all the
variables.]

b) The acceleration of a falling bouncy ball is due to gravity.
This acceleration is constant, so the motion is not SHM. [1 mark].

2) a) Maximum velocity = $(2\pi f)A = 2\pi \times 1.5 \times 0.05 = 0.47 \ ms^{-1}$
[1 mark].

b) Stopclock started when object released, so $x = A\cos(2\pi ft)$
[1 mark].
$x = 0.05 \times \cos(2\pi \times 1.5 \times 0.1) = 0.05 \times \cos(0.94) = 0.029 \ m$
[1 mark].

c) $x = A\cos(2\pi ft) \Rightarrow 0.01 = 0.05 \times \cos(2\pi \times 1.5t)$.
So $0.2 = \cos(3\pi t) \Rightarrow \cos^{-1}(0.2) = 3\pi t$. $3\pi t = 1.37 \Rightarrow t = 0.15 \ s$.
[1 mark for working, 1 mark for correct answer]

Don't forget to put your calculator in radian mode when you're solving
questions on circular motion — it's an easy mistake to make.

A2 Answers

Page 99 — Free and Forced Vibrations

1) a) When a system is forced to vibrate at a frequency that's close to, or the same as its natural frequency [1 mark] and oscillates with a much larger than usual amplitude [1 mark].

   b) See graph below. [1 mark] for showing a peak at the natural frequency, [1 mark] for a sharp peak.

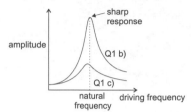

   c) See graph. [1 mark] for a smaller peak at the natural frequency [the peak will actually be slightly to the left of the natural frequency due to the damping, but you'll get the mark if the peak is at the same frequency in the diagram].

2) a) A system is critically damped if it returns to rest in the shortest time possible [1 mark] when it's displaced from equilibrium and released.

   b) e.g. suspension in a car [1 mark].

Unit 4: Section 3 — Thermal Physics

Page 101 — Solids, Liquids and Gases

1) Electrical energy supplied:
   $\Delta E = VI\Delta t = 12 \times 7.5 \times 180 = 16200$ J [1 mark]
   The temperature rise is $12.7 - 4.5 = 8.2$ °C

   Specific heat capacity: $c = \dfrac{\Delta E}{m\Delta \theta}$ [1 mark]

   $= \dfrac{16200}{2 \times 8.2} = 988$ J kg^{-1}°C^{-1} [1 mark]

   You need the right unit for the third mark — J kg^{-1} K^{-1} would be right too.

2) a) A molecule's energy is the sum of its potential and kinetic energy [1 mark]. The water is at 373 K so the molecules have the same kinetic energy [1 mark]. This means that one must have more potential energy than the other — i.e. one is liquid and the other is gas [1 mark].

   b) Total amount of energy needed to boil all the water:
   $\Delta E = ml = 2.26 \times 10^6 \times 0.5 = 1.13 \times 10^6$ J [1 mark]
   3 kW means you get 3000 J in a second, so
   time in seconds $= 1.13 \times 10^6 / 3000$ [1 mark] $= 377$ s [1 mark]

Page 103 — Ideal Gases

1) a) i) Number of moles $= \dfrac{\text{mass of gas}}{\text{molar mass}} = \dfrac{0.014}{0.028} = 0.5$ [1 mark]

   ii) Number of molecules = number of moles × Avogadro's constant
   $= 0.5 \times 6.02 \times 10^{23} = 3.01 \times 10^{23}$ [1 mark]

   b) $pV = nRT$, so $p = \dfrac{nRT}{V}$ [1 mark]

   $p = \dfrac{0.5 \times 8.31 \times 300}{0.01} = 125\,000$ Pa [1 mark]

   c) The pressure would also halve [1 mark] (because it is proportional to the number of molecules — $pV = NkT$).

2) $\dfrac{pV}{T} = $ constant

   At ground level, $\dfrac{pV}{T} = \dfrac{1 \times 10^5 \times 10}{293} = 3410$ JK^{-1} [1 mark]

   Higher up, pV/T will equal this same constant. [1 mark]
   So higher up, $p = \dfrac{\text{constant} \times T}{V} = \dfrac{3410 \times 260}{25}$

   $p = 35\,500$ Pa [1 mark]

Page 105 — The Pressure of an Ideal Gas

1) a) For example, put some smoke into a glass cell and shine a beam of light onto it [1 mark]. Use a microscope to view the smoke particles [1 mark].

   b) In Brownian motion, particles continually change direction so must be acted on by an external force [1 mark]. The nature of this force is uneven and random, which is consistent with a force caused by collisions between randomly moving particles [1 mark].

   c) The random movement of the particles causes them to collide with the container and exert an outward force [1 mark].

Page 107 — Internal Energy and Temperature

1) a) 6.02×10^{23} molecules [1 mark]

   b) $E = \dfrac{3}{2}kT$ [1 mark] $= [3 \times (1.38 \times 10^{-23}) \times 300] \div 2$

   $= 6.21 \times 10^{-21}$ J [1 mark]

   c) The nitrogen molecules are constantly colliding and transferring energy between themselves, so have different energies [1 mark].

2) i) The average kinetic energy would halve because it is proportional to the absolute temperature [1 mark].

   ii) Internal energy is the sum of the kinetic energy of the particles so this would also halve [1 mark].

   iii) The mass of a molecule would not change because it is a physical property of the molecule and not dependent on temperature [1 mark]

   iv) Pressure is proportional to temperature, so the pressure would halve [1 mark].

A2 Answers

Page 109 — Electric Fields

1)a)

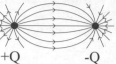

+Q   -Q

Recognisable pattern around the charges (not just in between)
[1 mark], lines equally spaced around the charges and joined to the
charges, and general symmetry of the diagram [1 mark], arrows
along field lines between the charges with arrows pointing away
from the positive and towards the negative charge [1 mark].

b) $E = \dfrac{Q}{4\pi\varepsilon_0 r^2}$ *[1 mark]*

$E = \dfrac{2 \times 1.6 \times 10^{-19}}{4\pi \times 8.85 \times 10^{-12} \times \left(3.5 \times 10^{-10}\right)^2}$ *[1 mark]*

$= 2.3 \times 10^{10}\ Vm^{-1} or\ NC^{-1}$ *[1 mark]*

2)a) $E = V/d = 1500/(4.5 \times 10^{-3}) = 3.3 \times 10^5$ *[1 mark]* Vm^{-1} *[1 mark]*
The field is perpendicular to the plates.   *[1 mark]*

b) $d = 2 \times (4.5 \times 10^{-3}) = 9.0 \times 10^{-3}\ m$ *[1 mark]*
$E = V/d \Rightarrow V = Ed = [1500/(4.5 \times 10^{-3})] \times 9 \times 10^{-3} = 3000\ V$
[1 mark]

Page 111 — Magnetic Fields

1)a) $F = BIl = 2 \times 10^{-5} \times 3 \times 0.04$ *[1 mark]* $= 2.4 \times 10^{-6}\ N$ *[1 mark]*
b) $F = BIl\sin\theta = 2.4 \times 10^{-6} \times \sin 30° = 2.4 \times 10^{-6} \times 0.5$ *[1 mark]*
$= 1.2 \times 10^{-6}\ N$ *[1 mark]*

Page 113 — Charged Particles in Magnetic Fields

1)a) $F = Bqv = 0.77 \times 1.6 \times 10^{-19} \times 5 \times 10^6$ *[1 mark]*
$= 6.16 \times 10^{-13}\ N$ *[1 mark]*

b) *The force acting on the electron is always at right angles to its*
velocity and the speed of the electron is constant. This is the
condition for circular motion. [1 mark]

2) *Electromagnetic force = centripetal force [1 mark]*
so, $Bqv = mv^2 / r$ *[1 mark]*

so, $r = \dfrac{mv}{Bq} = \dfrac{9.11 \times 10^{-31} \times 2.3 \times 10^7}{0.6 \times 10^{-3} \times 1.6 \times 10^{-19}} = 0.218\ m$ *[1 mark]*

3) $r = \dfrac{mv}{Bq}$ *[1 mark], which rearranges to give* $B = \dfrac{mv}{rq}$.

v, r *and* q *are all constant, so* $\dfrac{B}{m} = constant$ *[1 mark]*
Find the constant when $B = 0.20\ T$, $m = 35$: $0.20 \div 35 = 5.7 \times 10^{-3}$.
Now use this value to find B *when* $m = 37$:
$B = 37 \times 5.7 \times 10^{-3} = 0.21\ 1$ *[1 mark]*

Page 115 — Electromagnetic Induction

1)a) $\phi = BA$ *[1 mark]* $= 2 \times 10^{-3} \times 0.23 = 4.6 \times 10^{-4}\ Wb$ *[1 mark]*
b) $\Phi = BAN$ *[1 mark]* $= 2 \times 10^{-3} \times 0.23 \times 150 = 0.069\ Wb$
[1 mark]

c) $V = \dfrac{\Delta\Phi}{\Delta t} = \dfrac{(B_{start} - B_{end})AN}{\Delta t}$

$= \dfrac{(2 \times 10^{-3} - 1.5 \times 10^{-3})(0.23 \times 150)}{2.5} = 6.9 \times 10^{-3}\ V$

[3 marks available, one for each stage of the workings]

2)a) $\Phi = BAN$ *[1 mark]*
$= 0.9 \times 0.01 \times 500 = 4.5\ Wb$ *[1 mark]*

b) *Find the flux linkage after the movement:*
$\Phi = N\phi = NBA \cos\theta$ *[1 mark]*
$= 500 \times 0.9 \times 0.01 \times \cos(90) = 0\ Wb$ *[1 mark]*

$V = \dfrac{\Delta\Phi}{\Delta t}$ *[1 mark]*

$= \dfrac{4.5 - 0}{0.5} = 9\ V$ *[1 mark]*

Page 117 — Electromagnetic Induction

1)a) $V = Blv$ *[1 mark]* $= 60 \times 10^{-6} \times 30 \times 100 = 0.18\ V$ *[1 mark]*

b) *[1 mark]*

resistance

2)a) $\dfrac{V_p}{V_s} = \dfrac{N_p}{N_s}$ *[1 mark] so,* $N_s = \dfrac{45 \times 150}{9} = 750$ turns   *[1 mark]*

b) $\dfrac{V_s}{V_p} = \dfrac{N_s}{N_p}$ *[1 mark] so,* $V_s = \dfrac{V_p N_s}{N_p} = \dfrac{9 \times 90}{150} = 5.4\ V$ *[1 mark]*

c) *A transformer that increases the voltage [1 mark].*

Unit 5: Section 2 — Capacitors and Exponential Decay

Page 119 — Capacitors

1)a) *Capacitors are used to control a camera's flash by providing a short*
pulse of high current. [1 mark]
b) *Capacitors are suitable for this because they can deliver a short*
pulse of high current [1 mark], which results in a brief flash of
bright light when needed [1 mark].

2) *Capacitance* $= \dfrac{Q}{V} =$ *gradient of line* $= \dfrac{660\ \mu C}{3\ V} = 220\ \mu F.$

[1 mark for 'gradient', 1 mark for correct answer.]
Charge stored $= Q =$ *area* $= 15 \times 10^{-6} \times 66 = 990\ \mu C.$
[1 mark for 'area', 1 mark for correct answer.]

3)a) $W = \dfrac{1}{2}CV^2$ *[1 mark]* $= \dfrac{1}{2} \times 0.5 \times 12^2 = 36\ J$ *[1 mark]*

b) $Q = CV$ *[1 mark]* $= 0.5 \times 12 = 6\ C$ *[1 mark]*

A2 Answers

4) The voltage across all the components in a parallel circuit is the same as the source voltage, whereas in a series circuit the source voltage is shared between the components [1 mark]. This means that the two capacitors in parallel will each have a higher voltage across them than the two in series [1 mark]. Charge stored is proportional to voltage, so the capacitors in parallel will store more charge than those in series [1 mark]. [1 mark for a clear sequence of ideas.]

Page 121 — Charging and Discharging

1) a) The charge falls to 37% after **CR** seconds [1 mark], so $t = 1000 \times 2.5 \times 10^{-4} = 0.25$ seconds [1 mark]

b) $Q = Q_0 e^{-\frac{t}{CR}}$ [1 mark], so after 0.7 seconds:

$Q = Q_0 e^{-\frac{0.7}{0.25}} = Q_0 \times 0.06$ [1 mark]. There is 6% of the initial charge left on the capacitor after 0.7 seconds [1 mark].

c) i) The total charge stored will double [1 mark].
   ii) None [1 mark].
   iii) None [1 mark].

Unit 5: Section 3 — Nuclear Physics

Page 123 — Scattering to Determine Structure

1) a) The nuclear model states that an atom consists of a central mass with a positive charge (the nucleus) [1 mark], surrounded by orbiting negative electrons [1 mark]. The nucleus makes up a tiny proportion of the volume of an atom, but most of its mass [1 mark].

b) For alpha particles to pass straight through the foil, there must be a relatively large amount of empty space in an atom [1 mark]. For the alpha particles to be deflected at such large angles, the atom must contain a region of concentrated mass, greater than that of an alpha particle — the nucleus [1 mark]. This region must have the same charge as an alpha particle (positive) in order to deflect the particles [1 mark].

2) 8 protons, 8 electrons and 8 neutrons [2 marks for all correct, 1 mark for two correct]

Page 125 — Nuclear Radius and Density

1) Isotopes are atoms with the same number of protons, but different numbers of neutrons. / Isotopes are atoms with the same proton numbers, but different nucleon numbers [1 mark]. Two isotopes will have the same chemical properties as this depends on the number of protons [1 mark], but may have different nuclear stabilities [1 mark].

2) Volume $= \frac{4}{3}\pi r^3 = \frac{4}{3}\pi \left(8.53 \times 10^{-15}\right)^3 = 2.6 \times 10^{-42}$ m^3 [1 mark]

So density $(\rho) = \frac{m}{V} = \frac{3.75 \times 10^{-25}}{2.6 \times 10^{-42}} = 1.44 \times 10^{17}$ kg m^{-3} [1 mark]

3) The density of a gold nucleus is much larger than the density of a gold atom [1 mark]. This implies that the majority of a gold atom's mass is contained in the nucleus [1 mark].
The nucleus is small compared to the size of the atom [1 mark]. There must be a lot of nearly empty space inside each atom [1 mark].

Page 127 — The Strong Nuclear Force

1) a)

$$F = \frac{1}{4\pi\varepsilon_o}\frac{Q_1 Q_2}{r^2} = \frac{1}{4\pi\left(8.85 \times 10^{-12}\right)}\frac{\left(1.6 \times 10^{-19}\right)\left(1.6 \times 10^{-19}\right)}{\left(9 \times 10^{-15}\right)^2} = 2.8\,N$$

[1 mark for working, 1 mark for correct answer]

b) The electrostatic force will increase [1 mark].

c) There is no electrostatic force between a proton and a neutron [1 mark] because a neutron has no charge [1 mark].

2) a) The strong interaction must be repulsive at very small nucleon separations to prevent the nucleus being crushed to a point [1 mark].

b) Beyond 10 fm, the strong interaction is smaller than the electrostatic force [1 mark]. This means the protons in the nucleus would be forced apart. So a nucleus bigger than this would be unstable. [1 mark]

Page 129 — Radioactive Emissions

1) Place different materials between the source and detector and measure the amount of radiation getting through [1 mark]:

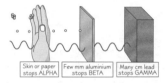

Skin or paper stops ALPHA   Few mm aluminium stops BETA   Many cm lead stops GAMMA

[1 mark for each material stopping correct radiation]

2) a) Radioactive isotopes in smoke alarms ionise the air between two electrodes [1 mark], which allows current to flow [1 mark]. If there's a fire, smoke absorbs the radiation, the current stops and the alarm sounds [1 mark].

b) Smoke alarms use sources of alpha radiation [1 mark]. Alpha radiation is sufficiently ionising to allow a current to flow between the electrodes [1 mark], but can only travel short distances so won't cause a hazard to people [1 mark].

Page 131— Exponential Law of Decay

1) Any one of: You can't say which atom/nucleus in a sample will be the next one to decay. / You can only estimate the fraction of nuclei that will decay or the probability an atom/nucleus will decay in a given time. / You cannot say exactly how many atoms will decay in a given time. [1 mark]

2) a) Activity, **A** = measured activity – background activity
= 750 – 50 = 700 Bq [1 mark]
$A = \lambda N \Rightarrow 700 = 50\,000\,\lambda$ [1 mark] So $\lambda = 0.014$ s^{-1} [1 mark]

b) $T_{\frac{1}{2}} = \frac{\ln 2}{\lambda} = \frac{0.693}{0.014} = 49.5$ seconds

[2 marks available — 1 mark for the half-life equation, 1 mark for the correct half-life]

c) $N = N_0 e^{-\lambda t} = 50\,000 \times e^{-0.014 \times 300} = 750$
[2 marks available — 1 mark for the decay equation, 1 mark for the number of atoms remaining after 300 seconds]

A2 Answers

Page 133 — Nuclear Decay

1)a) $^{226}_{88}Ra \rightarrow ^{222}_{86}Rn + ^{4}_{2}\alpha$ *[3 marks available — 1 mark for alpha particle, 1 mark each for proton and nucleon number of radon]*

b) $^{40}_{19}K \rightarrow ^{40}_{20}Ca + ^{0}_{-1}\beta$ *[3 marks available — 1 mark for beta particle, 1 mark each for proton and nucleon number of calcium]*

2) Mass defect = $(6.695 \times 10^{-27}) - (6.645 \times 10^{-27}) = 5.0 \times 10^{-29}$ kg *[1 mark]*. Using the equation $E = mc^2$ *[1 mark]*, $E = (5.0 \times 10^{-29}) \times (3 \times 10^8)^2 = 4.5 \times 10^{-12}$ J *[1 mark]*

Page 135 — Classification of Particles

1) Proton, electron and electron antineutrino. *[1 mark]*
The electron and the electron antineutrino are leptons. *[1 mark]*
Leptons are not affected by the strong interaction, so the decay can't be due to the strong interaction. *[1 mark]*

2) Mesons are hadrons but the muon is a lepton. *[1 mark]*
The muon is a fundamental particle but mesons are not. Mesons are built up from simpler particles. *[1 mark]*
Mesons feel the strong interaction but the muon does not. *[1 mark]*

Page 137 — Antiparticles

1) $e^+ + e^- \rightarrow \gamma + \gamma$ *[1 mark]*. This is called annihilation *[1 mark]*.

2) The protons, neutrons and electrons which make up the iron atoms would need to annihilate with their antiparticles *[1 mark]*. No antiparticles are available in the iron block *[1 mark]*.

3) The creation of a particle of matter requires the creation of its antiparticle. In this case no antineutron has been produced *[1 mark]*.
Also note that the total baryon number would have increased from 2 to 3 and that's not allowed.

Page 139 — Quarks

1) **uud**

2) $\pi^- = d\bar{u}$ *[1 mark]*
Charge of down quark = −1/3 unit.
Charge of anti-up quark = −2/3 unit
Total charge = −1 unit *[1 mark]*

3) The weak interaction converts a down quark into an up quark plus an electron and an electron antineutrino. *[1 mark]*
The neutron (udd) becomes a proton (uud). *[1 mark]*

4) The baryon number changes from 2 to 1 so baryon number is not conserved *[1 mark]*. The strangeness changes from 0 to 1 so strangeness is not conserved *[1 mark]*.

Page 141 — Binding Energy

1)a) There are 6 protons and 8 neutrons, so the mass of individual parts
= $(6 \times 1.007276) + (8 \times 1.008665) = 14.112976$ u *[1 mark]*
Mass of $^{14}_{6}C$ nucleus = 13.999948 u
so, mass defect = 14.112976 − 13.999948 = 0.113028 u *[1 mark]*
Converting this into kg gives mass defect = 1.88×10^{-28} kg *[1 mark]*

b) $E = mc^2 = (1.88 \times 10^{-28}) \times (3 \times 10^8)^2 = 1.69 \times 10^{-11}$ J *[1 mark]*
1 MeV = 1.6×10^{-13} J, so energy = $\frac{1.69 \times 10^{-11}}{1.6 \times 10^{-13}} = 106$ MeV *[1 mark]*

2)a) Fusion *[1 mark]*

b) The increase in binding energy per nucleon is about 0.86 MeV *[1 mark]*. There are 2 nucleons in ^{2}H, so the increase in binding energy is about 1.72 MeV — so about 1.7 MeV is released (ignoring the positron) *[1 mark]*.

Page 143 — Nuclear Fission and Fusion

1)a) Nuclear fission can be induced by neutrons and produces more neutrons during the process *[1 mark]*. This means that each fission reaction induces more fission reactions to occur, resulting in a ongoing chain of reactions *[1 mark]*.

b) For example, control rods limit the rate of fission by absorbing neutrons *[1 mark]*. The number of neutrons absorbed by the rods is controlled by varying the amount they are inserted into the reactor *[1 mark]*. A suitable material for the control rods is boron *[1 mark]*.

c) In an emergency shut-down, the control rods are released into the reactor *[1 mark]*. The control rods absorb the neutrons, and stop the reaction as quickly as possible *[1 mark]*.

2) Advantages: keeping the reaction in the nuclear reactor going doesn't produce any waste gases that could be harmful to the environment, e.g. sulfur dioxide (leading to acid rain) or carbon dioxide *[1 mark]*.
It can be used to supply a continuous supply of electricity, unlike some renewable sources *[1 mark]*.
Disadvantages (any two of): problems with the reactor getting out of control / risks of radiation from radioactive waste produced / the emissions released in the case of an accident / the long half-life of nuclear waste.
[1 mark for each disadvantage, maximum 2 marks]

Unit 5: Section 4 — Medical Imaging

Page 145 — X-Ray Imaging

1)a) Half-value thickness is the thickness of material required to reduce the intensity of an X-ray beam to half its original value *[1 mark]*.

b) $\mu = \frac{\ln 2}{x_{\frac{1}{2}}} = \frac{\ln 2}{3} = 0.23$ mm^{-1} *[1 mark]*, $I = I_0 e^{-\mu x} \Rightarrow \frac{I}{I_0} = e^{-\mu x}$ *[1 mark]*.
So, $0.01 = e^{-0.23x} \Rightarrow \ln(0.01) = -0.23x$ *[1 mark]*, $x = 20$mm *[1 mark]*.

A2 Answers

Page 147 — Ultrasound Imaging

1) a) $\alpha = \left(\dfrac{Z_{tissue} - Z_{air}}{Z_{tissue} + Z_{air}}\right)^2 = \left(\dfrac{1630 - 0.430}{1630 + 0.430}\right)^2$ [1 mark], $\alpha = 0.999$ [1 mark]

b) From part a), 0.1% enters the body when no gel is used [1 mark].

$\alpha = \left(\dfrac{Z_{tissue} - Z_{gel}}{Z_{tissue} + Z_{gel}}\right)^2 = \left(\dfrac{1630 - 1500}{1630 + 1500}\right)^2 = 0.002$ [1 mark], so 99.8% of the ultrasound is transmitted [1 mark]. Ratio is ~1000 [1 mark].

2) a) $Z = \rho v$ [1 mark], $v = (1.63 \times 10^6)/(1.09 \times 10^3)$
   $= 1495\ ms^{-1} \approx 1.50\ kms^{-1}$ [1 mark].

b) A pulse from the far side of the head travels an extra 2d cm, where d is the diameter of the head [1 mark].
Time taken to travel this distance = $2.4 \times 50 = 120\ \mu s$ [1 mark].
Distance = speed $\times$ time, so $2d = 1500 \times 120 \times 10^{-6}$
$= 0.18\ m$ [1 mark]. So $d = 9\ cm$ [1 mark].

Page 149 — Magnetic Resonance Imaging

1) a)

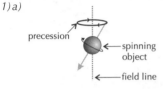

precession — spinning object — field line

The axis of rotation of a spinning object [1 mark] describes a circular path around the field line [1 mark].

b) The Larmor frequency is the frequency at which the protons, or other particles, in a magnetic field precess [1 mark]. In an MRI scanner it is typically 10^3–10^9 Hz [1 mark].

2) The patient lies in the centre of a large magnet, which produces a magnetic field [1 mark]. The magnetic field aligns hydrogen protons in the patient's body [1 mark]. Radio frequency coils are used to transmit radio waves, which cause the aligned protons to absorb energy [1 mark]. When the radio waves stop the protons emit the stored energy as radio waves, which are recorded by the scanner [1 mark]. A computer analyses the received radio waves to produce an image of the patient's body [1 mark].

3) Advantages, e.g. no known side effects / doesn't use ionising radiation / an image can be made for any slice in any orientation of the body / images of soft tissues are higher quality than using other techniques (e.g. CT, X-ray) / contrast can be weighted to investigate different situations.
Disadvantages, e.g. other techniques give better quality images of bony structures / people can suffer claustrophobia inside the scanner / MRI cannot be used on people with pacemakers/some metal implants / MRI scanners are very expensive.
[1 mark for each advantage explained, to a maximum of 3 marks. 1 mark for each disadvantage explained, to a maximum of 3 marks. 6 marks are available in total.]

Page 151 — Medical Uses of Nuclear Radiation

1) The patient is injected with a medical tracer consisting of a gamma source/positron-emitter bound to a substance used by the body [1 mark]. After a period of time, the radiation in the patient's body is recorded using a gamma camera/PET scanner [1 mark]. A computer uses this information to form an image, which might show a tumour as an area of high metabolic activity [1 mark].

2) a) The ionising energy of the particles released in 1000 s will be:
$1000 \times 3 \times 10^{10} \times 8 \times 10^{-13} = 24\ J$ [1 mark]
Absorbed dose = ionising energy $\div$ mass = $24 \div 70 = 0.34\ Gy$
[1 mark]

b) The effective dose of this absorbed dose is: $H = 0.343 \times 20$
$= 6.9\ Sv$ [1 mark]

Unit 5: Section 5 — Modelling the Universe

Page 153 — The Solar System & Astronomical Distances

1) [1 mark each for 5 sensible points], e.g., planets are generally much bigger than comets. Planets are made out of rock and gas whereas comets are largely made of frozen substances like ice. Planets have almost circular orbits whereas comets have highly elliptical orbits. Comets have tails when they are close to the Sun. Comets can take millions of years to orbit the Sun; planets have much shorter periods.

2) a) A light-year is the distance travelled by a photon of light through a vacuum in one year [1 mark].

b) Seconds in a year = $365.25 \times 24 \times 60 \times 60 = 3.16 \times 10^7\ s$ [1 mark].
Distance = $c \times$ time = $3.0 \times 10^8 \times 3.16 \times 10^7 = 9.5 \times 10^{15}\ m$
[1 mark].

c) To see something, light must reach us. Light travels at a finite speed, so it takes time for that to happen [1 mark]. The further out we see, the further back in time we're looking. The Universe is ~14 billion years old so we can't see further than ~14 billion light years. [1 mark].

Page 155 — Stellar Evolution

1) High mass stars spend much less time on the main sequence than low mass stars [1 mark]. As a red giant, low mass stars can only fuse hydrogen and/or helium but the highest mass stars can fuse nuclei up to iron [1 mark]. Lower mass stars eject their atmospheres to become white dwarfs [1 mark], but high mass stars explode in supernovae [1 mark] to leave neutron stars [1 mark] or black holes [1 mark].

Page 157 — The Big Bang Model of the Universe

1) a) $v = H_0 d$ [1 mark] where v is recessional velocity (in kms^{-1}), d is distance (in Mpc) and H_0 is Hubble's constant (in $kms^{-1}Mpc^{-1}$) [1 mark]

b) Hubble's law suggests that the Universe originated with the Big Bang [1 mark] and has been expanding ever since. [1 mark]

c) i) $H_0 = v \div d = 50\ kms^{-1} \div 1\ Mpc^{-1}$.
$50\ kms^{-1} = 50 \times 10^3\ ms^{-1}$ and $1\ Mpc^{-1} = 3.09 \times 10^{22}\ m$
So, $H_0 = 50 \times 10^3\ ms^{-1} \div 3.09 \times 10^{22}\ m = 1.62 \times 10^{-18}\ s^{-1}$
[1 mark for the correct value, 1 mark for the correct unit]

ii) $t = 1/H_0$ [1 mark]
$t = 1/1.62 \times 10^{-18} = 6.18 \times 10^{17}\ s \approx 20$ billion years [1 mark]
The observable Universe has a radius of 20 billion light years.
[1 mark]

A2 Answers

2) a) $z \approx v/c$ [1 mark] so $v \approx 0.37 \times 3.0 \times 10^8 \approx 1.1 \times 10^8$ ms^{-1} [1 mark]

b) $d = v/H_0 \approx 1.1 \times 10^8 / 2.4 \times 10^{-18} = 4.6 \times 10^{25}$ m [1 mark]
$= 4.6 \times 10^{25} / 9.5 \times 10^{15}$ ly $= 4.9$ billion ly [1 mark]

3) The cosmic background radiation is microwave radiation [1 mark] showing a continuous spectrum [1 mark] of a temperature of about 3 K [1 mark]. It is isotropic and homogeneous [1 mark].
It suggests that the ancient Universe was very hot, producing lots of electromagnetic radiation [1 mark] and that its expansion has stretched the radiation into the microwave region [1 mark].
[1 mark for clearly linking the evidence to the explanation.]

Page 159 — Evolution of the Universe

1) a) The amount of curvature of the Universe depends on its average density [1 mark]. Current estimates for the density of the Universe are close to the critical density required for the Universe to be flat [1 mark].

b) A flat Universe can be modelled by a 2-dimensional plane where parallel lines never meet and the angles in a triangle add up to 180° [1 mark]. In a **flat** Universe, gravity is **just strong enough** to stop the expansion at $t = \infty$ [1 mark], which means that the Universe will expand for ever, but more and more slowly with time [1 mark].

2) a) Find H_0 in SI units: $H_0 = 100$ $kms^{-1}Mpc^{-1}$
So, $H_0 = 100 \times 10^3 \div 3.09 \times 10^{22} = 3.236 \times 10^{-18}$ s^{-1} [1 mark]

Find the average density using: $\rho_0 = \dfrac{3H_0^{\,2}}{8\pi G}$ [1 mark]
and $G = 6.67 \times 10^{-11}$ Nm^2kg^{-2} [1 mark].

$\rho_0 = \dfrac{3 \times \left(3.236 \times 10^{-18}\right)^2}{8 \times \pi \times 6.67 \times 10^{-11}} = 1.87 \times 10^{-26}$ kgm^{-3} [1 mark]

b) The average density of the Universe is 1.87×10^{-26} kgm^{-3}, so on average, each m^3 of the Universe has a mass of 1.87×10^{-26} kg [1 mark].
This is equivalent to $1.87 \times 10^{-26} \div 1.7 \times 10^{-27} = 11$ hydrogen atoms. So there would be 11 hydrogen atoms in every m^3 if the entire mass of the Universe was hydrogen [1 mark].

3) A slight excess of matter was produced over antimatter [1 mark]. The matter and antimatter annihilated into photons [1 mark]. The remaining matter was in the form of quarks and leptons [1 mark]. As the Universe expanded and cooled the quarks combined to form particles like protons and neutrons [1 mark]. Some of the protons fused together to form helium [1 mark]. After about 300 000 years the Universe was cool enough for electrons to combine with the protons and helium nuclei to form atoms [1 mark]. The Universe became transparent since the photons could no longer interact with any free charges [1 mark]. There were fluctuations in the density of the Universe [1 mark] that allowed gravity to clump matter together, condensing matter into stars and galaxies [1 mark]. [1 mark for a clear, logical progression of ideas in the answer.]

Index

Index

Index

Index